BODYOLOGY

Published by Canbury Press, 2018

Canbury Press,
Kingston upon Thames, Surrey
www.canburypress.com

Cover: Ruth Blackford

Printed and bound in Great Britain by Clays Ltd, St Ives plc

ISBN
9780995497863 Paperback
9780995497870 Ebook
9780995497887 Audiobook

BODYOLOGY

The Curious Science of Our Bodies

Contents

What's it like to be struck by lightning

■ Charlotte Huff

Sometimes they'll keep the clothing, the strips of shirt or trousers that weren't cut away and discarded by the doctors and nurses. They'll tell and retell their story at family gatherings and online, sharing pictures and news reports of survivals like their own or far bigger tragedies. The video of a tourist hit on a Brazilian beach or the Texan struck dead while out running. The 65 people killed during four stormy days in Bangladesh.

Only by piecing together the bystander reports, the singed clothing and the burnt skin can survivors start to construct their own picture of the possible trajectory of the electrical current, one that can approach 200 million volts and travel at one-third of the speed of light.

In this way, Jaime Santana's family have stitched together some of what happened that Saturday afternoon in April 2016, through his injuries, burnt clothing and, most of all, his shredded broad-brimmed straw hat. 'It looks like somebody threw a cannonball through it,' says Sydney Vail, a trauma surgeon in Phoenix, Arizona, who helped care for Jaime after he arrived by ambulance, his heart having been shocked several times along the way as paramedics struggled to stabilise its rhythm.

Jaime had been horse-riding with his brother-in-law and two others in the mountains behind his brother-in-law's home outside Phoenix, a frequent weekend pastime. Dark clouds had formed, heading in their direction, so the group had started back.

They had nearly reached the house when it happened, says Alejandro Torres, Jaime's brother-in-law. He paces out the area involved, the landscape dotted with small creosote bushes just behind his acre of property. In the distance, the desert mountains rise, rippled chocolate-brown peaks against the horizon.

The riders had witnessed quite a bit of lightning as they neared Alejandro's house, enough that they had commented on the dramatic zigzags across the sky. But scarcely a drop of rain had fallen as they approached the horse corrals, just several hundred feet from the back of the property.

Alejandro doesn't think he was knocked out for long. When he regained consciousness, he was lying face down on the ground, sore all over. His horse was gone.

The two other riders appeared shaken but unharmed. Alejandro went looking for Jaime, who he found on the other side of his fallen horse. Alejandro brushed against the horse's legs as he walked passed. They felt hard, like metal, he says, punctuating his English with some Spanish. He reached Jaime: 'I see smoke coming up – that's when I got scared.' Flames were coming off of Jaime's chest. Three times Alejandro beat out the flames with his hands. Three times they reignited.

It wasn't until later, after a neighbour had come running from a distant property to help and the paramedics had arrived, that they began to realise what had happened – Jaime had been struck by lightning.

§ § §

Justin Gauger wishes his memory of when he was struck – while fishing for trout at a lake near Flagstaff, Arizona – wasn't so vivid. If it weren't, he wonders, perhaps the anxiety and lingering effects of post-traumatic stress disorder wouldn't have trailed him for so long. Even now, some three years later, when a storm moves in, the flickering flashes of light approaching, he's most comfortable sitting in his bathroom closet, monitoring its progress with an app on his phone.

An avid fisherman, Justin had initially been elated when the rain started that August afternoon. The storm had kicked up suddenly, as they often do during the summer monsoon season. Fish are more likely to bite when it's raining, he told his wife, Rachel.

But as the rain picked up, becoming stronger and then turning into hail, his wife and daughter headed for the truck, followed later by his son. The pellets grew larger, approaching golf ball size, and really started to hurt as they pounded Justin's head and body.

Giving up, he grabbed a nearby folding canvas chair – the charring on one corner is still visible today – and turned to head for the truck. Rachel was filming the storm from the front seat, planning to catch her husband streaking back as the hail intensified. She pulls up the video on her phone.

Initially all that's visible on the screen is white, a blur of hail hitting the windshield. Then a flash flickers across the screen, the only one that Rachel saw that day, the one that she believes felled her husband.

A crashing boom. A jolting, excruciating pain. 'My whole body was just stopped – I couldn't move any more,' Justin recalls. 'The pain was... I can't explain the pain except to say if you've ever put your finger in a light socket as a kid, multiply that feeling by a gazillion throughout your entire body. And I saw a white light surrounding my body – it was like I was in a bubble. Everything was slow motion. I felt like I was in a bubble for ever.'

A couple huddling under a nearby tree ran to Justin's assistance. They later told him that he was still clutching the chair. His body was smoking.

When Justin came to, he was looking up at people staring down, his ears ringing. Then he realised that he was paralysed from the waist down. 'Once I figured out that I couldn't move my legs, I started freaking out.'

Describing that day, sitting on his sofa at home, Justin draws one hand across his back, tracing the path of his burns, which at one point covered roughly a third of his body. They began near his right shoulder and extended diagonally across his torso, he says, and then continued along the outside of each leg.

He leaves and returns holding his hiking boots, tipping them to show several burn marks on the interior. Those dark roundish spots line up with the singed areas on the socks he was wearing and with the coin-sized burns he had on both feet, which were deep enough that he could put the tip of his finger inside.

The singed markings also align with several needle-sized holes located just above the thick rubber soles of his size 13 boots. Justin's best guess – based on reports from the nearby couple, along with the wound on his right shoulder – is that the lightning hit his upper body and then exited through his feet.

Although survivors frequently talk about entry and exit wounds, it's difficult to figure out in retrospect precisely what path the lightning took, says Mary Ann Cooper, a retired Chicago emergency physician and long-time lightning researcher. The visible evidence of lightning's wrath is more reflective, Cooper says, of the type of clothing a survivor had on, the coins they were carrying in their pockets and the jewellery they were wearing as the lightning flashed over them.

Lightning is responsible for more than 4,000 deaths worldwide annually – according to those documented in reports from 26 countries. (The true scope of lightning's casualties in the more impoverished and lightning-prone areas of the world, such as central Africa, is still being calculated.) Cooper is one of a small global cadre of doctors, meteorologists, electrical engineers and others who are driven to better understand how lightning injures people, and ideally how to avoid it in the first place.

Of every ten people hit by lightning, nine will survive to tell the tale. But they could suffer a variety of short- and long-term effects. The list is lengthy and daunting: cardiac arrest, confusion, seizures, dizziness, muscle aches, deafness, headaches, memory deficits, distractibility, personality changes and chronic pain, among others.

Many survivors have a story that they want to share. In postings online and during annual gatherings of Lightning Strike & Electric Shock Survivors International, they swap tales of their brush with nature's brutal force. The group has convened in the mountains of the south-eastern US every spring since its first meeting was held by 13 survivors in the early 1990s. In those pre-internet days, it was far more difficult to meet other survivors coping with the headaches, memory troubles, insomnia and other effects of a lightning strike, says Steve Marshburn, the group's founder, who has been living with symptoms since he was struck near a bank teller's window in 1969.

For nearly 30 years, he and his wife have run the organisation – which now has nearly 2,000 members – from their North Carolina home. They nearly cancelled this year's conference, as Marshburn, who is 72 years old, has been having some health issues. But the members wouldn't allow it, he says, a bit proudly.

The changes in personality and mood that survivors experience, sometimes with severe bouts of depression as well, can strain families and marriages, sometimes to breaking point. Cooper likes to use the analogy that lightning rewires the brain in much the same way that an electrical shock can scramble a computer – the exterior appears unharmed, but the software within that controls its functioning is damaged.

Both Marshburn and Cooper credit the organisation's very existence with saving lives, with it preventing at least 22 suicides according to Marshburn. It's not unusual for him to field a call in the middle of the night and talk for hours with someone in dire straits. He is drained afterwards, unable to do much for the next few days.

Cooper, who has attended some of these gatherings, has learned to hang back as survivors and their loved ones describe their symptoms. 'I still don't understand all of them,' she says. 'A lot of times I can't understand what's going on with these people. And I listen and I listen and I listen.'

Despite a deep vein of sympathy for survivors, some symptoms still strain Cooper's credulity. Some people maintain that they can detect a storm brewing long before it appears on the horizon. That's possible, Cooper says, given their heightened sensitivity to stormy signs in the wake of their trauma. She's less open to other reports – those who say that their computer freezes when they enter a room, or that the batteries in their garage door opener or other devices drain more quickly.

Yet, even after decades of research, Cooper and other lightning experts readily admit that there are many unresolved questions, in a field where there's little to no research funding to decipher the answers. It's not clear, for example, why some people appear to suffer seizure-related symptoms after their lightning injury. Also, are lightning survivors more vulnerable to other health problems, such as heart conditions, later in life?

Some survivors report feeling like medical nomads, as they struggle to find a doctor with even a passing familiarity with lightning-related injuries. Justin, who could move his legs within five hours of being struck, finally sought out help and related testing last year at the Mayo Clinic for his cognitive frustrations.

Along with coping with post-traumatic stress disorder, Justin chafes at living with a brain that doesn't function as fluidly as it once did. He doesn't see how he could possibly return to the type of work he used to shoulder, leading a small team that presented legal cases and helped defend the county against property value disputes. Talking on the phone one day, sounding quite articulate, he tries to convey the struggles lurking just beneath. 'My words in my head are jumbled. When I think about what I'm trying to say, it's all jumbled up. So when it comes out, it may not sound all right.'

§ § §

When someone is hit by lightning, it happens so fast that only a very tiny amount of electricity ricochets through the body. The vast majority travels around the outside in a 'flashover' effect, Cooper explains.

By way of comparison, coming into contact with high-voltage electricity, such as a downed wire, has the potential to cause more internal injuries, since the exposure can be more prolonged. A 'long' exposure might still be relatively brief – just a few seconds. But that's sufficient time for the electricity to penetrate the skin's surface, risking internal injuries, even to the point of cooking muscle and tissue to the extent that a hand or limb might need to be amputated.

So what causes external burns? Cooper explains that, as lightning flashes over the body, it might come into contact with sweat or raindrops on the skin's surface. Liquid water increases in volume when it's turned into steam, so even a small amount can create a 'vapour explosion'. 'It literally explodes the clothes off,' says Cooper. Sometimes the shoes too. However, shoes are more likely to be torn or damaged on the inside, because that's where the heat build-up and vapour explosion occurs. 'That's it,' Cooper responds when she's told about the singed markings on Justin's hiking boots.

As for clothing, steam will interact with it differently depending upon what it's made of. A leather jacket can trap the steam inside, burning the survivor's skin. Polyester can melt with just a few pieces left behind, primarily the stitching that once held together the seams of a shirt or a jacket that's no longer there, says Cooper, who has seen a decent quantity of post-lightning relics through the years.

Along with the burn marks visible on Jaime Santana's clothes, the cellphone he was carrying in his pocket melted, bonding to his pants. (His sister, Sara, now wishes that they had kept the phone but they tossed it, fearful that it carried some residual lightning current – a bit paranoid, she now realises.) While Jaime's family believes that lightning shredded his hat, causing it to expand upward and outward, Cooper is more dubious when she sees a photograph. There's no visible singeing, she notes. And the chunk of straw could have been lost during Jaime's tumble from the horse.

Cooper authored one of the first studies looking at lightning injuries, published nearly four decades ago, in which she reviewed 66 physician reports about seriously injured patients, including eight that she'd treated herself. Loss of consciousness was common. About one-third experienced at least some temporary paralysis in their arms or legs.

Those rates might be on the high side – Cooper points out that not all lightning patients are sufficiently injured that doctors write about their cases. But survivors do often describe temporary paralysis, like Justin suffered, or a loss of consciousness, although why it occurs is not clear.

More is understood about lightning's ability to scramble the electrical impulses of the heart, thanks to experiments with Australian sheep. Lightning's massive electrical current can temporarily stun the heart, says Chris Andrews, a physician and lightning researcher at the University of Queensland in Australia. Thankfully, though, the heart possesses a natural pacemaker. Frequently, it can reset itself.

The problem is that lightning can also knock out the region of the brain that controls breathing. This doesn't have a built-in reset, meaning a person's oxygen supply can become dangerously depleted. The risk then is that the heart will succumb to a second and potentially deadly arrest, Andrews says. 'If someone has lived to say, 'Yes, I was stunned [by lightning],' it's probable that their respiration wasn't completely wiped out, and re-established in time to keep the heart going.'

Andrews is well suited to conducting lightning studies, having trained both as an electrical engineer and as a physician. His research, looking at the impact of electrical current on sheep, is frequently credited with demonstrating how lightning's flashover current can still inflict damage within the body. One reason sheep were chosen, Andrews says, is that they're relatively close to humans in size. Another advantage is that the specific breed chosen, the barefaced Leicester, doesn't grow much wool around its head, making it similar to a human's.

During his studies, Andrews shocked anesthetised sheep with voltage levels roughly similar to a small lightning strike and photographed the electricity's path. He showed that as lightning flashes over, the electrical current enters critical portals into the body: the eyes, the ears, the mouth. This helps explain why damage to the eyes and ears is frequently reported by survivors. They might develop cataracts. Or their hearing can be permanently damaged, even after the initial post-boom ringing stops.

Particularly worrisome is that, by penetrating the ears, lightning can rapidly reach the brain region that controls breathing, Andrews says. Upon entering the body, the electricity can hitch a ride elsewhere, through the blood or the fluid surrounding the brain and the spinal cord. Once it reaches the bloodstream, Andrews says, the passage to the heart is very quick.

In Arizona, Jaime Santana survived the immediate lightning strike. The family's beloved horse Pelucha – from the Spanish for 'stuffed animal' – did not. One possibility, the trauma surgeon Sydney Vail and others speculate, is that the 1,500-pound steed absorbed a good portion of the lightning that nearly killed his 31-year-old rider.

Another reason Jaime survived is that, when he was struck, the neighbour who came running – someone who the family had never met before – immediately started CPR, and continued until the paramedics arrived. At one point, Alejandro says, one of the paramedics asked the other if they should stop, as Jaime wasn't responding. The neighbour insisted that they continue.

That CPR occurred immediately is 'the only reason he's alive,' says Vail. The neighbour later told the family that he had performed CPR 'hundreds and hundreds of times' in nearly two decades as a volunteer paramedic, says Jaime's sister,

Sara, her voice cracking as she talks. Before Jaime, no one had survived.

§§§

Lightning begins high up in the clouds, sometimes 15,000 to 25,000 feet above the earth's surface. As it descends toward the ground, the electricity is searching, searching, searching for something to connect with. It steps, almost stair-like, in a rapid-fire series of roughly 50-metre increments. Once lightning is 50 metres or so from the ground, it searches again pendulum-style in a nearby radius for 'the most convenient thing to hit the fastest,' says Ron Holle, a US meteorologist and long-time lightning researcher.

Prime candidates include isolated and pointed objects: trees, utility poles, buildings and occasionally people. The entire cloud-to-ground sequence happens blindingly fast.

The popular perception is that the chance of being struck by lightning is one in a million. There's some truth here, based on US data, if one only looks at deaths and injuries in a single year. But Holle, who believes that statistic is misleading, set out to crunch some other numbers. If someone lives until 80, their lifetime vulnerability increases to 1 in 13,000. Then consider that every victim knows at least ten people well, such as the friends and family of Jaime and Justin. Thus, any individual's lifetime probability of being personally affected by a lightning strike is even higher, a 1 in 1,300 chance.

Holle doesn't even like the word 'struck', saying it implies that lightning strikes hit the body directly. In fact, direct strikes are surprisingly rare. Holle, Cooper and several other prominent lightning researchers recently pooled their expertise and calculated that they're responsible for no more than 3 to 5 per cent of injuries. (Still, Vail, the trauma surgeon, surmises that Jaime was directly hit, given that he was riding in the desert with no trees or other tall objects nearby.)

Justin believes that he experienced what's called a side flash or side splash, in which the lightning 'splashes' from something that has been struck – such as a tree or telephone pole – hopscotching to a nearby object or person. Considered the second most common lightning hazard, side splashes inflict 20 to 30 per cent of injuries and fatalities.

By far the most common cause of injury is ground current, in which the electricity courses along the earth's surface, ensnaring within its circuitry a herd of cows or a group of people sleeping beneath a tent or a grass-thatched hut.

As a general rule, in high-income regions of the world men are more likely than women to be injured or killed by lightning; at least two-thirds of the time they're the victims, and possibly higher depending upon the study. One possibility is the propensity for 'men taking chances,' Holle quips, as well as work-related exposure. They are more likely to be on the younger side, in their 20s or 30s, and doing something outside, frequently on the water or nearby.

But what should you do if you find yourself stranded a long way from a building or car when a storm kicks up? Some guidance is available: avoid mountain peaks, tall trees or any body of water. Look for a ravine or a depression. Spread out your group, with at least 20 feet between each person, to reduce the risk of multiple injuries. Don't lie down, which boosts your exposure to ground current. There's even a recommended lightning position: crouched down, keeping the feet close together.

Still, don't dare to ask Holle about any of these suggestions. There's no such thing as a lightning-proof guarantee, he repeats more than once. 'There are cases where every one of these [strategies] has led to death.' In his cubicle at the control centre of the US National Lightning Detection Network (NLDN) in Tucson – operated by Vaisala, a Finland-based environmental observation company – Holle has accumulated stacks and stacks of folders filled with articles and other write-ups detailing a seemingly endless litany of lightning-related scenarios involving people or animals. Deaths and injuries that have occurred in tents, or during sports competitions, or to individuals huddled beneath a golf shelter or a picnic shelter or some other type of shelter.

That word whitewashes the reality, Holle says, as so-called 'shelters' can become 'death traps' during a lightning storm. They provide protection from getting wet – that's it.

On a series of large screens lining two walls of a room at NLDN's offices in Tucson, Holle can see where cloud-to-ground lightning is flashing in real time, picked up by strategically positioned sensors in the US and elsewhere. Satellite data has shown that certain regions of the world, generally those near the equator, are lightning-dense. Venezuela, Colombia, the Democratic Republic of the Congo and Pakistan all rank among the top ten lightning hotspots.

Initially, lightning safety campaigns promoted the 30/30 rule, which relied upon individuals counting off the seconds after lightning flashed. If thunder rumbled before they reached 30, lightning was close enough to pose a threat.

But there's been a move away from that advice for various reasons, Holle says. One is practical: it's not always easy to figure out which rumble of thunder corresponds to which lightning flash.

Instead, for simplicity's sake, everyone from schoolchildren to their grandparents these days is advised: 'When thunder roars, go indoors.'

§ § §

Better education isn't the only reason why lightning deaths have steadily declined in the US, Australia and other high-income regions. Housing construction has improved. Jobs have moved indoors. In the US alone, annual fatalities have fallen from more than 450 in the early 1990s to fewer than 50 in recent years.

There's always room for improvement, though. Arizona, for example, ranks high in the US when looking at lightning deaths per state population. Holle's theory is that people stay outside longer in the desert as the rain isn't necessarily heavy during storms. That's why casualties can occur, even before the storm arrives, with people dallying their way to shelter while lightning stretches out in front of the dark clouds.

Still, people in high-income countries have it easy, compared to those in regions where people have no choice but to work outside in all conditions and lightning-safe buildings are scarce. In one analysis of agricultural-related lightning deaths outside of the US, Holle learned that more than half of them occurred in India, followed by Bangladesh and the Philippines. The victims were young (early 20s for the men, early 30s for the women) and were working in farms and paddy fields.

Cooper was hit full-force with the emotional impact of what lightning can do in Africa when she attended a 2011 lightning conference in Nepal. The presenters were arranged in alphabetical order by country, so Cooper, by then retired as an emergency physician but still doing lightning-related work, was sat between the presenters from Uganda and Zambia. Richard Tushemereirwe, the Ugandan representative, kept fussing with his slides while waiting to present.

'When he got up to give his presentation, he was almost in tears,' she recalls. 'He said, 'I found out from my research that we had 75 people die in Uganda during the last lightning season." And just that summer, he related, 18 students had died in a single lightning strike to a school in central Uganda.

In an email, Tushemereirwe described how the lightning protection that some schools do install can create a false sense of security. A rod may be installed on the roofline of one school building. But it's not grounded. Even worse, local residents might believe that the single rod also protects nearby buildings, wrote Tushemereirwe, who serves as senior science adviser to Uganda's president.

Nor does home provide a sanctuary when lightning laces the sky, as housing in rural regions of Africa is frequently constructed from mud and grass. Thus, the mantra 'When thunder roars, go indoors' is essentially useless, Cooper notes with considerable frustration. Families are at risk 24/7.

Lightning deaths go unreported or are missed entirely. It might appear, for instance, that a fire killed an entire family. But that assumption misses a key piece of the tragedy. Sometimes it's lightning that sets the grass roof ablaze, temporarily paralysing the family members within, so they're unable to escape the flames.

On a bus trip to a banquet after Tushemereirwe's presentation, he and Cooper fell into talking. It was a discussion that led to a collaboration and, in 2014, the creation of a non-profit organisation now called the African Centres for Lightning and Electromagnetics Network, with Cooper its founding director. Zambia was the second country to join after Uganda. Leaders of several others have expressed interest, Cooper says.

The organisation is trying to develop a cellphone alert system so that fishermen and others in the Lake Victoria region can report severe weather heading their way. They are starting to educate school teachers about lightning safety and are setting up graduate study programmes.

Another priority is Ugandan schools, frequently the most substantial structures in a given community. The first lightning protection system was installed in a school in late 2016, as were two more earlier this year. Keeping the focus on protecting children, it's been learned through other lightning safety efforts, gets adults' attention, Cooper says. Adults the world over believe they are immune, she states flatly. 'But if you tell them that their kids are going to get injured, they pay attention.'

Still, making headway has been an uphill climb, slowed by fundraising and installation logistics. Cooper sound-ed a bit weary and discouraged after her most recent trip to Uganda this spring. The country has thousands of vulnerable schools. She's now searching for deeper pockets through foundation or governmental funding. 'We've protected three of them. Oh my God, how will we ever be able to,' she says, her voice trailing off. 'It's so overwhelming, I just want to quit. I don't see how we are ever going to be able to impact this.'

§§§

The rain that had threatened all afternoon didn't start to fall until Sara and Alejandro were driving to Maricopa Medical Center in Phoenix. Alejandro sat tense, holding on to his terrible knowledge. 'All of this way, I was thinking, 'He's dead. How do I tell her?''

When they arrived, Alejandro was stunned to learn that Jaime was in surgery. Surgery? There was still hope.

Jaime had arrived at the Phoenix trauma centre with an abnormal heart rhythm, bleeding in the brain, bruising to the lungs and damage to other organs, including his liver, according to Vail. Second- and third-degree burns covered nearly one-fifth of his body. Doctors put him into a chemically induced coma for nearly two weeks to allow his body to recover, a ventilator helping him breathe.

Jaime finally returned home after five months of treatment and rehabilitation, which is continuing. 'The hardest part for me is that I can't walk,' he says from the living room of his parents' house. The doctors have described some of Jaime's nerves as still 'dormant', says his sister, Sara, something that they hope time and rehabilitation will mend.

'We're living through something that we never thought in a million years would happen,' says Lucia, Jaime's mother, reflecting on the strike and Jaime's miraculous survival, Sara translating. They've stopped asking why lightning caught him in its crosshairs that April afternoon. 'We're never going to be able to answer why,' Sara says. So now it's time for Jaime to start thinking about 'what's next' with the new life he's been given. The family is planning a party, with a mariachi band, to celebrate Jaime's first year of life moving forward.

When Sara and Alejandro returned home from the hospital the day after the strike, Alejandro called to his wife from the backyard. On the railing of the round pen where they work the horses, adjacent to their corrals, a peacock was perched, his colourful feathers flowing behind.

Outside of a zoo they had never seen a peacock in Arizona before. They kept the peacock and later found it a mate. Now a family of peacocks fills one of the corral stalls. When Sara looked up what the striking bird symbolises, the answers scrolled back, catching her breath: renewal, resurrection, immortality.

This story was first published on 23 May 2017
by Wellcome on mosaicscience.com.

How and why we colour hair

■ Rebecca Guenard

Every two months Barclay Cunningham goes through a process that begins with taking an antihistamine tablet. After a few hours, she smears a thick layer of antihistamine cream across her forehead, around her ears and over her neck. Finally, she shields the area with ripped-up plastic carrier bags.

All this so she can dye her hair.

It didn't start out this bad. Cunningham coloured her hair for a decade without any problems. Then, one day, she noticed that the skin on her ears was inflamed after she'd dyed her hair. She fashioned plastic bag earmuffs and carried on colouring. But the allergic reaction persisted, so her precautions became more elaborate. Now if she dyes her hair without these measures, she gets an itchy, blistery, pus-filled rash that lasts for weeks.

Suffering for the sake of tinted tresses is not a modern-day phenomenon. Humans have dyed for thousands of years, experimenting with ever-changing, often vicious, formulas to achieve a new hair colour.

The chemical history of modern hair dyes reveals that, while they were once part of an innovative industry, progress has stalled, and today they rely on antiquated methods. But consumers are not exactly pressuring the industry to innovate. Not when they are so desperate to change their hair colour that they're willing to discreetly pick scabs from their hair, as Cunningham does, for weeks after colouring.

Aesthetic tendencies drift with marketing and cultural currents, but our drive to alter ourselves is a constant. As anthropologist Harry Shapiro wrote: 'So universal is this urge to improve on nature...that one is almost tempted to regard it as an instinct.'

§ § §

Hundreds of plastic practice mannequins, lips pursed in proper model pouts, float around the halls of the Energizing Summit, an annual event of the American Board of Certified Haircolorists. You don't really ever adjust to seeing the disembodied heads, be they upside down in clear plastic bags (the handle cinched tight around the neck for easier carrying), gazing out of boxes in the hotel lobby, or mounted on poles, like some kind of punishment from Tudor England.

Hairdressers from around the USA, all with stunning hair colour and impeccably maintained roots, criss-cross the poorly lit basement of the Marriott Hotel at Los Angeles Airport. They're here for two days of sessions dedicated to the science of dyeing hair.

Right away I realise that I have a lot to learn. Hair colourists, it seems, speak a different language to the rest of us. They talk of 'volume' (concentration) and 'lift' (lightening). And it turns out I have been making a faux pas. 'We dye Easter eggs,' one Summit instructor gently informs me. 'We colour hair.'

But after a day and a half, I am still waiting for some science. Then I find Tom Despenza. He has years of experience working in research and development at Clairol – a career that began when, as a microbiology student, his car broke down in front of a beauty school. He is now retired and owns his own hair colour company called Chromatics.

When I catch up with Tom at the Summit, he has been teaching his popular class 'Forget the Hype! Let's get real', which dispels the years of hearsay that makes up the beauty school curriculum.

Understanding the dyes used on hair is not as simple as understanding the colour wheel. As we all learned in art class, any colour can be obtained by mixing the three primary colours of red, yellow and blue. If you want orange, you mix yellow and red; if you want purple, you combine red and blue; and if you want brown, you mix all three.

Beauticians are taught the same thing when it comes to hair – that brown dye is a combination of three different dyes. 'That's just fictitious information,' says Despenza. 'Brown hair colour is made up of two chemicals.' Both chemicals are colourless, he explains, but they produce brown through a chemical reaction that occurs when they're combined.

An important distinction exists between colour and dye. Hairdressers are not applying pigments (at least not in the case of permanent hair dye), they are applying a mixture of chemicals to initiate dye formation. The individual dye molecules have to be linked together before they emit colour, so dyes have to sit on the head for 30 minutes to allow this reaction to occur.

§ § §

In the mid-1800s, English chemist William Henry Perkin serendipitously synthesised the first non-natural dye: starting with coal tar, he was hoping to produce the malaria drug quinine but instead created mauve. His discovery revolutionised the textile industry and launched the petrochemical industry. Natural dyes just didn't have the staying power and vivid colours of the dye Perkin created. Never before had such a steadfast dye been found.

Soon after, August Hofmann (Perkin's chemistry professor) noticed that a dye he had derived from coal tar formed a colour when exposed to air. The molecule responsible was para-phenylenediamine, or PPD, the foundation of most permanent hair dyes today.

Although hair is a protein fibre, like wool, the dyeing process for textiles cannot be duplicated on the head. To get wool to take a dye, you must boil the wool in an acidic solution for an hour. The equivalent for hair is to bathe it in the chemical ammonia. Ammonia separates the protective protein layers, allowing dye compounds to penetrate the hair shaft and access the underlying pigment, melanin.

Melanin is what gives colour to human skin, eyes and hair. It's the ratio of two types of melanin – eumelanin and pheomelanin – that determines your natural hair colour. And it's the size and shape that the melanin molecules form when they cluster in the hair shaft that gives the unique tones within a hair colour. For example, blondes and brunettes have about the same ratio of eumelanin molecules to pheomelanin molecules, but blondes have fewer molecules overall. Natural blond hair also contains smaller melanin clusters, which reflect light more than the larger clusters found in dark hair.

Along with ammonia, hair dye formulas contain hydrogen peroxide, a bleaching agent. Peroxide serves two purposes: it reacts with the melanin in hair, extinguishing its natural colour, and provokes a reaction between PPD molecules. The trapped colour-emitting molecule will remain in the hair, too big to escape, and the natural colour will appear only as the hair grows out.

Early on, dye chemists realised that if they added a secondary molecule, called a coupler, they could manipulate the chemicals – a carbon here, a couple of nitrogens there – and multiply the colour choices that were available with PPD alone. Different methods have been proposed, but beauty manufacturers have yet to accept a permanent hair colour formula without PPD or its related compound p-aminophenol.

§ § §

For 125 years, the oxidative reaction of PPD has been the extent of hair dye technology. Dr David Lewis, emeritus professor at the University of Leeds in the UK, thinks that this is 'crazy'. 'Now, I know a lot about dyes and dye stuffs in the textile industry. We would never dream of using this on textiles,' he says. 'Primitive, archaic, all these things come to mind. Why do they persist on putting it on human heads?'

As a research professor, Lewis acted as a consultant for cosmetics companies, but he always felt uncomfortable about their insistence upon using the same old oxidative formulas. Lewis retired from academia ten years ago to launch Green Chemicals, a company that aims to develop safer consumer goods. His company introduced a more environmentally friendly flame retardant, and now Lewis wants to overhaul hair dyes.

One issue is how dyes work: Lewis says that the colour molecules become electron scavengers along the way to creating beautiful brown tresses. This need for electrons is not fulfilled exclusively by other dye molecules, so the electron scavengers also aggressively pursue the skin – causing allergic reactions and potentially damaging DNA.

Lewis is also worried that the beauty industry has too much power over consumer safety. The modern era of the Food and Drug Administration (FDA) began in 1906, when it was known as the Bureau of Chemistry. In 1930 it adopted the name we know today. The FDA has banned many types of dyes since, but it has always officially deemed coal tar dyes safe, especially for hair colouring, as long as consumers were warned of the possibility of skin irritation. To this day, coal tar dyes (which are now derived from petroleum) do not require FDA certification.

In 1979 the FDA tried to insist that hair dye manufacturers place the following label on their products: 'Warning – Contains an ingredient that can penetrate your skin and has been determined to cause cancer in laboratory animals.' The ingredient referred to is 4-MMPD, 4-methoxy-m-phenylenediamine, a dye with a structure very similar to PPD that, according to the FDA, showed sufficient scientific evidence of being carcinogenic. Manufacturers disagreed and threatened to sue the FDA if they pressed for the label. The FDA backed down. A few years later, manufacturers removed the carcinogenic compound from their formulas, while maintaining that 4-MMPD was safe.

There is some research into the potential risk of dyes. In 2001, researchers at the University of Southern California published a paper in the International Journal of Cancer concluding that women who frequently dye their hair were twice as likely to develop bladder cancer than those who abstain. The European Commission on Consumer Safety took note. A panel of scientists evaluated the paper, deemed it scientifically credible and recommended that the EU reassess hair dye regulations.

Over the past decade the Science Committee on Consumer Products (SCCP) – a committee of the European Commission mandated to assess and report on product safety – has collected and evaluated manufacturers' data and published opinions on a number of hair dye ingredients. This re-evaluation of hair colour ingredients by the EU has highlighted two issues.

The first is that sensitisation to dye chemicals has grown considerably. The EU has categorised 27 hair colour ingredients as sensitisers, listing 10 of them as extreme and 13 strong. Although the first exposure to a sensitiser might have no noticeable effect, a subsequent exposure – to the same chemical or to similar chemicals in temporary tattoos or textiles, for example – could lead to an allergic reaction. In the worst case, it could trigger anaphylaxis, an extreme and potentially fatal allergic response.

The second issue is a lack of data on what dye chemicals do inside the human body. When in doubt, the European Commission bans the use of a particular chemical. In 2006, then-European Commission Vice-President Günter Verheugen said in a press release: 'Substances for which there is no proof that they are safe will disappear from the market. Our high safety standards do not only protect EU consumers, they also give legal certainty to European cosmetics industry.' It has prohibited 22 hair dye chemicals so far – and more are likely to be added to the list, which is updated annually. Most recently the SCCP deemed 2-chloro-p-phenylenediamine, used to colour eyebrows and lashes, unsafe on the grounds of insufficient toxicology data.

When the SCCP released the findings on sensitivity in early 2007, Colipa (the European cosmetic trade association, now known as Cosmetics Europe) published a statement to 'reinforce its confidence in the safety of hair dyes'. Although stating their support of the European Commission's ongoing work to evaluate the safety of hair dyes, they argued that the dye chemicals were being tested in isolation and that the findings did not give an indication of the health risks the chemicals could pose if used in consumer products as instructed.

Scientists working for the industry continue to point out that no undisputed epidemiological studies show a significant risk of cancer among people who colour their hair. Unless you look at a population that is exposed to hair dye every day: hairdressers. Hairdressers have a 5 per cent greater chance of contracting bladder cancer than the general population.

It struck me that there was no mention of the safety of hair dye chemicals during any of the instructional classes I attended at the Energizing Summit. When I overheard a student being advised to think about her long-term health as a hairdresser, I looked up to see whether it related to contact with dyes (studies have shown that wearing gloves greatly reduces the amount of dye compounds absorbed into the body). But it turned out that the student was being counselled on her wrist position, not the use of gloves.

§§§

In the 1970s, anthropologist Justine Cordwell wrote a paper entitled 'The very human arts of transformation'. In it, she wrote: 'The anthropological analysis of clothing and adornment should be based on the assumption that mankind, from earliest times, has probably regarded the human body as the primary form of sculpture – and not been particularly pleased with what he has seen.'

Indeed, archaeological evidence shows that the use of dyes by humans dates back to the Palaeolithic period. Early humans used the iron oxide contained in dirt to decorate their dwellings, textiles and bodies with the colour red. It wasn't too long until they applied the dyes to their heads.

Ancient Egyptians dyed their hair, but rarely did so while it was on their heads. They shaved it off, then curled and braided it to fashion wigs to protect their bald heads from the sun. Black was the most popular colour until around the 12th century BCE, when plant material was used to colour the wigs red, blue or green, and gold powder was used to create yellow.

Of natural dyes, henna endures. The ancients also used saffron, indigo and alfalfa. But natural dyes only coat the hair temporarily, and people wanted chemically altered tresses. Analysing hair samples has revealed that the Greeks and Romans used permanent black hair dye thousands of years ago. They mixed substances that we know today as lead oxide and calcium hydroxide to create a lead sulphide nanoparticle, which forms when the chemicals interact with sulphur linkages in keratin, a protein in hair. When the direct application of lead proved too toxic, the Romans changed their black dye formula to one made by fermenting leeches for two months in a lead vessel.

Prostitutes during the early years of the Roman Empire were required to have yellow hair to indicate their profession. Most wore wigs, but some soaked their hair in a solution made from the ashes of burnt plants or nuts to achieve the colour chemically. Meanwhile, Germans coloured their hair red by applying a mixture of beechwood ash and goats' fat.

With the development of the scientific method in the early modern period, dyers took a more analytical approach to changing hair colour, testing the efficacy and safety of new formulas. *Delights for Ladies*, a recipe book of household essentials published in the early 1600s, recommends using Oyle of Vitrioll to colour black hair chestnut. The book cautions to avoid touching the skin – sound advice given that today we know Oyle of Vitrioll as sulphuric acid.

The fashion for Italian blondes repeated itself – as hair colour trends do – several hundred years later when, in the 1700s, Venetian women would recline in the sun on specially built terraces with their hair drenched in corrosive solutions of lye to achieve golden locks. Blond hair was no longer limited to prostitutes.

Yet dyes were used for more than fashion or to signify occupation. Cordwell identifies several instances where hair colour was changed for other reasons; for example, Afghans believed that dyeing their hair red with henna could cure a bad headache.

§§§

Beauty is a multi-billion-pound industry that's continuing to grow. According to, cosmetics manufacturing will have brought in $255 billion (£155 billion) in revenue globally in 2014. The industry remained stable through the recession and, as incomes increase with recovery, demand for high-priced beauty products means that global profits are estimated to increase to $316 billion by 2019.

Globally, haircare products are the largest portion of the beauty industry and secure nearly a quarter of industry revenue. In the USA, within hair and nail salons, hair-colouring services account for 18 per cent of revenue. An estimated 70 per cent of women in the USA use hair colouring products.

Reflecting on the heritage of hair dyes, you can't help but ask: why do so many people still colour their hair? Why would someone go through the rigmarole and tolerate the expense, the itching and the smell? Whatever drives our desire to change the colour of our hair, one thing is certain: people have deep emotional ties to what covers their scalps.

This is clearly true for Barclay Cunningham. At just 12 years old, she began experimenting with her hair, using a spray-in hair-lightening chemical. As an adult, she searched for years for the right hair colour. 'Never once has it occurred to me to simply not dye my hair,' Barclay says. 'The 'me' of hair colour happens to come out of a box. The 'me' that grew out of my head was not right.'

This story was first published on 16 December 2014 by Wellcome on mosaicscience.com

The man with the golden blood

■ Penny Bailey

His doctor drove him over the border. It was quicker that way: if the man donated in Switzerland, his blood would be delayed while paperwork was filled out and authorisations sought.

The nurse in Annemasse, France, could tell from the label on the blood bag destined for Paris that this blood was pretty unusual. But when she read the details closely, her eyes widened. Surely it was impossible for this man seated beside her to be alive, let alone apparently healthy?

Thomas smiled to himself. Very few people in the world knew his blood type did – could – exist. And even fewer shared it. In 50 years, researchers have turned up only 40 or so other people on the planet with the same precious, life-saving blood in their veins.

§§§

Red blood cells carry oxygen to all the cells and tissues in our body. If we lose a lot of blood in surgery or an accident, we need more of it – fast. Hence the hundreds of millions of people flowing through blood donation centres across the world, and the thousands of vehicles transporting bags of blood to processing centres and hospitals.

It would be straightforward if we all had the same blood. But we don't. On the surface of every one of our red blood cells, we have up to 342 antigens – molecules capable of triggering the production of specialised proteins called antibodies. It is the presence or absence of particular antigens that determines someone's blood type.

Some 160 of the 342 blood group antigens are 'high-prevalence', which means that they are found on the red blood cells of most people. If you lack an antigen that 99 per cent of people in the world are positive for, then your blood is considered rare. If you lack one that 99.99 per cent of people are positive for, then you have very rare blood.

If a particular high-prevalence antigen is missing from your red blood cells, then you are 'negative' for that blood group. If you receive blood from a 'positive' donor, then your own antibodies may react with the incompatible donor blood cells, triggering a further response from the immune system. These transfusion reactions can be lethal.

Because so few people have it, by definition, rare blood is hardly ever needed. But when it is, finding a donor and getting the blood to a patient in crisis can become a desperate race against the clock. It will almost certainly involve a convoluted international network of people working invisibly behind the bustle of 'ordinary' blood donation to track down a donor in one country and fly a bag of their blood to another.

§ § §

Forty years ago, when ten-year-old Thomas went into the University Hospital of Geneva with a routine childhood infection, his blood test revealed something very curious: he appeared to be missing an entire blood group system.

There are 35 blood group systems, organised according to the genes that carry the information to produce the antigens within each system. The majority of the 342 blood group antigens belong to one of these systems. The Rh system (formerly known as 'Rhesus') is the largest, containing 61 antigens.

The most important of these Rh antigens, the D antigen, is quite often missing in Caucasians, of whom around 15 per cent are Rh D negative (more commonly, though inaccurately, known as Rh-negative blood). But Thomas seemed to be lacking all the Rh antigens. If this suspicion proved correct, it would make his blood type Rh_{null} – one of the rarest in the world, and a phenomenal discovery for the hospital haematologists.

Rh_{null} blood was first described in 1961, in an Aboriginal Australian woman. Until then, doctors had assumed that an embryo missing all Rh blood cell antigens would not survive, let alone grow into a normal, thriving adult. By 2010, nearly five decades later, some 43 people with Rh_{null} blood had been reported worldwide.

Hardly able to believe what she was seeing, Dr Marie-José Stelling, then head of the haematology and immunohaematology laboratory at the University Hospital of Geneva, sent Thomas' blood for analysis in Amsterdam and then in Paris. The results confirmed her findings: Thomas had Rh_{null} blood. And with that, he had instantly become infinitely precious to medicine and science.

Researchers seeking to unravel the mysteries of the physiological role of the intriguingly complex Rh system are keen to get hold of Rh_{null} blood, as it offers the perfect 'knockout' system. Rare negative blood is so sought after for research that even though all samples stored in blood banks are anonymised, there have been cases where scientists have tried to track down and approach individual donors directly to ask for blood.

And because Rh_{null} blood can be considered 'universal' blood for anyone with rare blood types within the Rh system, its life-saving capability is enormous. As such, it's also highly prized by doctors – although it will be given to patients only in extreme circumstances, and after very careful consideration, because it may be nigh on impossible to replace. 'It's the golden blood,' says Dr Thierry Peyrard, the current Director of the National Immunohematology Reference Laboratory in Paris.

Blood groups are inherited, and Rh_{null} is known to run in families. So the next step for the haematologists in Geneva was to test Thomas' family in the hope of finding another source, particularly as Thomas wouldn't be able to donate until he turned 18. Things looked even more hopeful when it turned out Thomas' grandfathers were third-degree cousins. But the tests showed Thomas' Rh_{null} blood was due to two completely different random mutations on both sides. Pure chance, twice over, in the face of vanishingly small odds.

§§§

In 2013, Walter Udoeyop received a letter from an old friend back in Nigeria. Father Gerald Anietie Akata's 70-year-old mother had a tumour in her heart, but no hospital in Nigeria could perform the surgery she needed. Akata enclosed his mother's medical records, asking for Walter's help.

Walter, a consultant at Johnson City Medical Center, Tennessee, knew from the start this wasn't going to be easy. Francisca Akata's operation would cost a daunting $150,000 minimum if she had it in the USA. Father Akata had been a pastor in Johnson City for several years, and Walter initially hoped to enlist the help of the church and hospitals his friend had served in. But neither could raise such a large amount of money. He recalled that another friend had recently had open-heart surgery in the United Arab Emirates (UAE) for only $20,000. He phoned the hospital there, and the staff agreed to operate on Francisca. Father Akata's parishioners in the USA and Nigeria raised the money, and three months later, Francisca Akata was flying eastwards towards the UAE.

But a few days after her admission, the doctors told Francisca that blood tests had revealed that she had a rare blood type, shared by 0.2 per cent of the white population: Lutheran b negative. To complicate the matter, she was also O negative – the uncommon, but not officially rare blood type that many of us have heard of, shared by around 5 per cent of people. The combination made Francisca's blood so rare it would be difficult, if not impossible, to find a match for her.

Since there was no compatible blood in the UAE or any of the other Gulf States, Mrs Akata had to fly back home and wait until matching blood was found. The hospital searched for blood but couldn't find any in the two weeks that followed.

Walter requested Francisca's blood tests from the hospital and then began the search for compatible blood in the USA. He tried blood centre after blood centre across the country before he was referred to the American Rare Donor Program in Philadelphia, a database of all rare blood donors in America. Finally, he had located some suitable donors.

His relief was short lived because, as Thomas or Peyrard could have told him, it's a lot harder for blood to cross borders than it is for people. 'You would not imagine how difficult it is when you have to import or export rare blood,' Peyrard says. 'Your patient is dying, and you have people in an office asking for this paper and that form. It's just crazy. It's not a TV set, it's not a car. It's blood.'

Sometimes sending blood from one country to another is more than a bureaucratic nightmare. As Walter now discovered, the hospital in the UAE had a policy not to accept blood donations from outside the Gulf States, which meant that Francisca wouldn't be able to use the blood he had found in America.

The situation looked bleak. But then, after a chance meeting with a colleague, Father Akata found out about a small general hospital in Cameroon, Nigeria's neighbour to the east, that had set up a heart surgery programme with funding from the Catholic Church. Walter got in touch with the surgeons there, who confirmed that they could do the surgery if he could supply compatible blood.

The American Rare Donor Program contacted the South African National Blood Service, which had four suitable donors listed; however, one of these was unreachable, one wasn't able to donate until later in the year, and two had been medically retired from donating. These are all common problems with rare donors. There are limits placed on how often people can donate. What's more, keeping track of donors can also be a challenge – some get ill or die, and others move home without updating the blood services.

There were two units of compatible blood in South Africa's frozen rare blood bank, but frozen blood has a 48-hour lifespan, compared to four weeks for fresh blood. If it got held up at customs, or delayed for any other reason, it would be unusable by the time it reached the hospital in Cameroon. To use the frozen South African blood, Francisca would have to have her operation in South Africa. Walter was running out of options.

§ § §

When he turned 18, Thomas was encouraged to donate blood for himself. There is now no frozen blood bank in Switzerland, so his blood is stored in the rare blood banks in Paris and Amsterdam. He travels to France to donate, avoiding the bureaucratic machinery that would grind into action if his blood had to be sent over the Swiss border to Paris.

The first urgent request came a few years after Thomas began donating, when he got a phone call asking if he would mind taking, and paying for, a taxi to the blood centre in Geneva to donate for a newborn baby. That moment brought it starkly home to him how valuable his blood was. It was perhaps also the first intimation that the costs of donating would ultimately be his. Some countries do pay donors (and some pay more for rare blood) to encourage donations. But the majority of national blood services don't pay, to deter donors with infections such as HIV.

The altruistic ethos of the blood services in western Europe means that they also don't cover donors' travel costs or time off work, both of which would constitute payment for blood. They can't even send a taxi to take a rare donor to a blood centre, although the blood itself can travel by taxi from the centre to a patient.

It's something that can become a frustrating and potentially fatal problem. Thierry Peyrard told me that he was recently contacted by a doctor in Zurich asking for blood with another rare combination of negatives, for a patient about to undergo surgery. There were only four compatible donors in France, and Peyrard could only contact two by phone. One of them was a 64-year-old lady living in a village near Toulouse. Although she was more than willing to go to the blood centre in the city and donate, since she can't drive she simply couldn't get there.

So unless his doctor drives him over the border again – a courtesy going far beyond the call of duty between a doctor and patient – Thomas will take a day's holiday from work, twice a year, and travel to Annemasse in France to donate, paying his own travel costs there and back.

§ § §

Walter had one last chance to find the blood Francisca Akata so desperately needed: a small laboratory on the other side of the Atlantic. The modest size of the International Blood Group Reference Laboratory (IBGRL) in Filton, near Bristol in England, is misleading; the expertise at the IBGRL means it's one of the world's leading laboratories in rare blood identification.

If the donor and recipient are in different countries, the blood services of both countries will negotiate costs. The requesting country usually covers the cost of flying the blood in at 4°C, the temperature at which fresh red blood cells have to be preserved to keep them intact before transfusion.

'It's generally a reciprocal agreement between countries so that no one who ever needs blood is penalised for being where they are,' Nicole Thornton, Head of Red Cell Reference at the IBGRL, told me. 'Some countries charge a bit more per unit of blood, but there's no hard and fast rule. Most don't charge too much because they might be in the same situation at some point.' In the UK the charge is £125.23 per bag of blood.

Walter contacted Thornton, who searched the International Rare Donor Panel and identified 550 active O negative/Lutheran b negative donors worldwide. Because they are inherited, blood types tend to pool in specific populations, and 400 of those donors turned out to be in the UK – most of them in or around London. A flight from the UK to Cameroon would be much shorter than a flight from the USA. And the blood would be fresh. As the blood was available, and because it wouldn't make logistical sense to fly a second consignment if Francisca needed more than four units, the UK blood services agreed to send six bags of blood. All six UK donors who received a phone call and were able to donate did so within days.

In Rochester, Kent, England, it was James'* 104th donation. He first donated as an 18-year-old in the army in 1957, when he was told his O negative blood was 'good for blue babies'. In 1985 he got a letter explaining his blood was rare (although not how rare) and asking if anyone in his family would donate so their blood could be tested. The family members he asked were reluctant, however, believing they 'needed their blood for themselves'.

A few years later, in the early 1990s, James got his first phone call from the blood services, asking whether he would mind 'popping down to the local hospital to donate blood for someone in Holland'. There, he was told that a taxi was waiting for his blood. He didn't think this particularly unusual, saying that it was all in a day's donation.

He wasn't surprised to be called up again, but he didn't know that his blood was going to Cameroon. And previously he hadn't known he was Lutheran b negative, as well as O negative. He was surprised and mildly intrigued to learn that there were only 550 known people with the same blood in the world.

The bag of James' blood joined the five bags from the other donors, and all six were couriered to Tooting in south London to start their 7000-km journey.

This was when Walter discovered how remote the hospital really was. Kumbo, in the mountains of north-west Cameroon, is more than 400 km north of both Douala and Yaoundé, the sites of the country's major international airports. The blood would have to travel for several hours to get there, bumping over a dirt road in the March heat. Even packed in ice, it would be hard to keep it at the cell-preserving 4°C.

Francisca Akata's blood landed at the international airport in Douala and was cleared through customs by noon on Friday 21 March 2014. A helicopter, supplied at the last minute by the hospital to avoid the lengthy road journey, was waiting outside. Her blood flew the rest of the way to Kumbo and arrived at 14.00, just after Francisca had been wheeled into the operating theatre.

The operation was a success, and the Catholic Diocese of Kumbo found her somewhere to recuperate before flying home. Walter still marvels at the efforts of so many people – on three separate continents – to save one life, 'just like the shepherd who left 99 sheep and went after the one that was lost.'

§ § §

Since his blood can be given to anyone with a negative Rh blood type, Thomas could save countless lives. But if he ever needs blood himself, he can receive only Rh_{null} blood. If he donates a unit for himself, he has to permit it to be used by anyone else who might need it.

This leaves Thomas dependent on other Rh_{null} donors. But of the other 40-odd people known worldwide with Rh_{null} blood, only six or so besides Thomas are thought to donate. And they're all a long way away: their locations include Brazil, Japan, China, the USA and Ireland. The reluctance to donate is perhaps understandable, but it places an added burden on the people who do give their blood. It's also probably why

Thomas, when Peyrard and I met him in Lausanne, greeted us with mild amusement. 'Is it interesting to put a face to the bag of blood?'

Over tea, he described the impact of his blood on his life. As a child he couldn't go to summer camp because his parents feared he might have an accident. As an adult he takes reasonable precautions: he drives carefully and doesn't travel to countries without modern hospitals. He keeps a card from the French National Immunohematology Reference Laboratory in Paris, confirming his Rh_{null} blood type, in his wallet in case he is ever hospitalised. But one thing that is in his blood – and that of almost everyone growing up in the shadow of the Alps – is skiing. Abstaining seems to have been an option he never even considered.

The only apparent health effect he experiences is mild anaemia, which is why he was advised to donate twice a year instead of four times. Intriguingly, one doctor once asked whether he has a quick temper. In fact, the opposite is the case: 'I am very calm. If it's just my personality or if my blood has an impact or not, I don't know.'

On the whole, Thomas is laid back about his 'condition'. 'I don't have a problem like haemophilia that has an impact on daily life. In that sense, I'm lucky. I'm glad also that when I was told I had this special blood, they told me it's okay to have children. I was allowed to have a family, so I'm happy.'

Donors like James and Thomas never learn what has happened to their blood – and haematologists don't fly across borders to express their appreciation. But on this day, as we sat in a room full of spring sunlight looking out at the hazy white-flecked peaks, Peyrard told Thomas that his blood had saved lives. Just recently, a unit was sent back from France to Switzerland for a young child who would otherwise have died.

On one level, Thomas' blood does divide him from the rest of us. On another, as the French philosopher Simone Weil observed, every separation is a link. And Thomas' different blood has given him the odd unexpected perk. When he was due for conscription, the doctor who first told him about his blood – Dr Marie-José Stelling – wrote to the army saying it was too dangerous for him to do military service, so he was exempted. Over the course of the past 40-odd years, Thomas and Stelling have developed a particularly close relationship. Last year, she joined him and his family and friends to celebrate his 50th birthday party on a boat on Lake Geneva. 'She was very kind,' Thomas says. 'She saw the human aspect of being Rh_{null} – not just a bag of blood.'

This story was first published on 21 October 2014 by Wellcome on mosaicscience.com

Why dieters can't rely on calories

■ Cynthia Graber and Nicola Twilley

'For me, a calorie is a unit of measurement that's a real pain in the rear.'

Bo Nash is 38. He lives in Arlington, Texas, where he's a technology director for a textbook publisher. And he's 5'10' and 245 lbs – which means he is classed as obese.

In an effort to lose weight, Nash uses an app to record the calories he consumes and a Fitbit band to track the energy he expends. These tools bring an apparent precision: Nash can quantify the calories in each cracker crunched and stair climbed. But when it comes to weight gain, he finds that not all calories are equal. How much weight he gains or loses seems to depend less on the total number of calories, and more on where the calories come from and how he consumes them. The unit, he says, has a 'nebulous quality to it'.

Tara Haelle is also obese. She had her second son on St Patrick's Day in 2014, and hasn't been able to lose the 70 lbs she gained during pregnancy. Haelle is a freelance science journalist, based in Illinois. She understands the science of weight loss, but, like Nash, doesn't see it translate into practice. 'It makes sense from a mathematical and scientific and even visceral level that what you put in and what you take out, measured in the discrete unit of the calorie, should balance,' says Haelle. 'But it doesn't seem to work that way.'

Nash and Haelle are in good company: more than two-thirds of American adults are overweight or obese. For many of them, the cure is diet: one in three are attempting to lose weight in this way at any given moment. Yet there is ample evidence that diets rarely lead to sustained weight loss. These are expensive failures. This inability to curb the extraordinary prevalence of obesity costs the United States more than $147 billion in healthcare, as well as $4.3 billion in job absenteeism and yet more in lost productivity.

At the heart of this issue is a single unit of measurement – the calorie – and some seemingly straightforward arithmetic. 'To lose weight, you must use up more calories than you take in,' according to the Centers for Disease Control and Prevention. Dieters like Nash and Haelle could eat all their meals at McDonald's and still lose weight, provided they burn enough calories, says Marion Nestle, professor of nutrition, food studies and public health at New York University. 'Really, that's all it takes.'

But Nash and Haelle do not find weight control so simple. And part of the problem goes way beyond individual self-control. The numbers logged in Nash's Fitbit, or printed on the food labels that Haelle reads religiously, are at best good guesses. Worse yet, as scientists are increasingly finding, some of those calorie counts are flat-out wrong – off by more than enough, for instance, to wipe out the calories Haelle burns by running an extra mile on a treadmill. A calorie isn't just a calorie. And our mistaken faith in the power of this seemingly simple measurement may be hindering the fight against obesity.

§ § §

The process of counting calories begins in an anonymous office block in Maryland. The building is home to the Beltsville Human Nutrition Research Center, a facility run by the US Department of Agriculture. When we visit, the kitchen staff are preparing dinner for people enrolled in a study. Plastic dinner trays are laid out with meatloaf, mashed potatoes, corn, brown bread, a chocolate-chip scone, vanilla yoghurt and a can of tomato juice. The staff weigh and bag each item, sometimes adding an extra two-centimetre sliver of bread to ensure a tray's contents add up to the exact calorie requirements of each participant. 'We actually get compliments about the food,' says David Baer, a supervisory research physiologist with the Department.

The work that Baer and colleagues do draws on centuries-old techniques. Nestle traces modern attempts to understand food and energy back to a French aristocrat and chemist named Antoine Lavoisier. In the early 1780s, Lavoisier developed a triple-walled metal canister large enough to house a guinea pig. Inside the walls was a layer of ice. Lavoisier knew how much energy was required to melt ice, so he could estimate the heat the animal emitted by measuring the amount of water that dripped from the canister. What Lavoisier didn't realise – and never had time to find out; he was put to the guillotine during the Revolution – was that measuring the heat emitted by his guinea pigs was a way to estimate the amount of energy they had extracted from the food they were digesting.

Until recently, the scientists at Beltsville used what was essentially a scaled-up version of Lavoisier's canister to estimate the energy used by humans: a small room in which a person could sleep, eat, excrete, and walk on a treadmill, while temperature sensors embedded in the walls measured the heat given off and thus the calories burned. (We now measure this energy in calories. Roughly speaking, one calorie is the heat required to raise the temperature of one kilogram of water by one degree Celsius.) Today, those 'direct-heat' calorimeters have largely been replaced by 'indirect-heat' systems, in which sensors measure oxygen intake and carbon dioxide exhalations. Scientists know how much energy is used during the metabolic processes that create the carbon dioxide we breathe out, so they can work backwards to deduce that, for example, a human who has exhaled 15 litres of carbon dioxide must have used 94 calories of energy.

The facility's three indirect calorimeters are down the halls from the research kitchen. 'They're basically nothing more than walk-in coolers, modified to allow people to live in here,' physiologist William Rumpler explains as he shows us around. Inside each white room, a single bed is folded up against the wall, alongside a toilet, sink, a small desk and chair, and a short treadmill. A couple of airlocks allow food, urine, faeces and blood samples to be passed back and forth. Apart from these reminders of the room's purpose, the vinyl-floored, fluorescent-lit units resemble a 1970s dorm room. Rumpler explains that subjects typically spend 24 to 48 hours inside the calorimeter, following a highly structured schedule. A notice pinned to the door outlines the protocol for the latest study:

6:00 to 6:45pm – Dinner,
11:00pm – Latest bedtime, mandatory lights out,
11:00pm to 6:30am – Sleep, remain
in bed even if not sleeping.

In between meals, blood tests and bowel movements, calorimeter residents are asked to walk on the treadmill at 3 miles per hour for 30 minutes. They fill the rest of the day with what Rumpler calls 'low activity'. 'We encourage people to bring knitting or books to read,' he says. 'If you give people free hand, you'll be surprised by what they'll do inside the chamber.' He tells us that one of his less cooperative subjects smuggled in a bag of M&Ms, and then gave himself away by dropping them on the floor.

Using a bank of screens just outside the rooms, Rumpler can monitor exactly how many calories each subject is burning at any moment. Over the years, he and his colleagues have aggregated these individual results to arrive at numbers for general use: how many calories a 120-lb woman burns while running at 4.0 miles an hour, say, or the calories a sedentary man in his 60s needs to consume every day. It's the averages derived from thousands of extremely precise measurements that provide the numbers in Bo Nash's movement tracker and help Tara Haelle set a daily calorie intake target that is based on her height and weight.

Measuring the calories in food itself relies on another modification of Lavoisier's device. In 1848, an Irish chemist called

Thomas Andrews realised that he could estimate calorie content by setting food on fire in a chamber and measuring the temperature change in the surrounding water. (Burning food is chemically similar to the ways in which our bodies break food down, despite being much faster and less controlled.) Versions of Andrews's 'bomb calorimeter' are used to measure the calories in food today. At the Beltsville centre, samples of the meatloaf, mashed potatoes and tomato juice have been incinerated in the lab's bomb calorimeter. 'We freeze-dry it, crush into a powder, and fire it,' says Baer.

Humans are not bomb calorimeters, of course, and we don't extract every calorie from the food we eat. This problem was addressed at the end of the 19th century, in one of the more epic experiments in the history of nutrition science. Wilbur Atwater, a Department of Agriculture scientist, began by measuring the calories contained in more than 4,000 foods. Then he fed those foods to volunteers and collected their faeces, which he incinerated in a bomb calorimeter. After subtracting the energy measured in the faeces from that in the food, he arrived at the Atwater values, numbers that represent the available energy in each gram of protein, carbohydrate and fat. These century-old figures remain the basis for today's standards. When Baer wants to know the calories per gram figure for that night's meatloaf, he corrects the bomb calorimeter results using Atwater values.

§§§

This entire enterprise, from the Beltsville facility to the numbers on the packets of the food we buy, creates an aura of scientific precision around the business of counting calories. That precision is illusory.

The trouble begins at source, with the lists compiled by Atwater and others. Companies are allowed to incinerate freeze-dried pellets of product in a bomb calorimeter to arrive at calorie counts, though most avoid that hassle, says Marion Nestle. Some use the data developed by Atwater in the late 1800s. But the Food and Drug Administration (FDA) also allows companies to use a modified set of values, published by the Department of Agriculture in 1955, that take into account our ability to digest different foods in different ways.

Atwater's numbers say that Tara Haelle can extract 8.9 calories per gram of fat in a plate of her favourite Tex-Mex refried beans; the modified table shows that, thanks to the indigestibility of some of the plant fibres in legumes, she only gets 8.3 calories per gram. Depending on the calorie-measuring method that a company chooses – the FDA allows two more variations on the theme, for a total of five – a given serving of spaghetti can contain from 200 to 210 calories. These uncertainties can add up. Haelle and Bo Nash might deny themselves a snack or sweat out another few floors on the StairMaster to make sure they don't go 100 calories over their daily limit. If the data in their calorie counts is wrong, they can go over regardless.

There's also the issue of serving size. After visiting over 40 US chain restaurants, including Olive Garden, Outback Steak House and PF Chang's China Bistro, Susan Roberts of Tufts University's nutrition research centre and colleagues discovered that a dish listed as having, say, 500 calories could contain 800 instead. The difference could easily have been caused, says Roberts, by local chefs heaping on extra french fries or pouring a dollop more sauce. It would be almost impossible for a calorie-counting dieter to accurately estimate their intake given this kind of variation.

Even if the calorie counts themselves were accurate, dieters like Haelle and Nash would have to contend with the significant variations between the total calories in the food and the amount our bodies extract. These variations, which scientists have only recently started to understand, go beyond the inaccuracies in the numbers on the back of food packaging. In fact, the new research calls into question the validity of nutrition science's core belief that a calorie is a calorie.

Using the Beltsville facilities, for instance, Baer and his colleagues found that our bodies sometimes extract fewer calories than the number listed on the label. Participants in their studies absorbed around a third fewer calories from almonds than the modified Atwater values suggest. For walnuts, the difference was 21 per cent. This is good news for someone who is counting calories and likes to snack on almonds or walnuts: he or she is absorbing far fewer calories than expected. The difference, Baer suspects, is due to the nuts' particular structure: 'All the nutrients – the fat and the protein and things like that – they're inside this plant cell wall.' Unless those walls are broken down – by processing, chewing or cooking – some of the calories remain off-limits to the body, and thus are excreted rather than absorbed.

Another striking insight came from an attempt to eat like a chimp. In the early 1970s, Richard Wrangham, an anthropologist at Harvard University and author of the book Catching

Fire: How cooking made us human, observed wild chimps in Africa. Wrangham attempted to follow the entirely raw diet he saw the animals eating, snacking only on fruit, seeds, leaves, and insects such as termites and army ants. 'I discovered that it left me incredibly hungry,' he says. 'And then I realised that every human eats their food cooked.'

Wrangham and his colleagues have since shown that cooking unlaces microscopic structures that bind energy in foods, reducing the work our gut would otherwise have to do. It effectively outsources digestion to ovens and frying pans. Wrangham found that mice fed raw peanuts, for instance, lost significantly more weight than mice fed the equivalent amount of roasted peanut butter. The same effect holds true for meat: there are many more usable calories in a burger than in steak tartare. Different cooking methods matter, too. In 2015, Sri Lankan scientists discovered that they could more than halve the available calories in rice by adding coconut oil during cooking and then cooling the rice in the refrigerator.

Wrangham's findings have significant consequences for dieters. If Nash likes his porterhouse steak bloody, for example, he will likely be consuming several hundred calories less than if he has it well-done. Yet the FDA's methods for creating a nutrition label do not for the most part account for the differences between raw and cooked food, or pureed versus whole, let alone the structure of plant versus animal cells. A steak is a steak, as far as the FDA is concerned.

Industrial food processing, which subjects foods to extremely high temperatures and pressures, might be freeing up even more calories. The food industry, says Wrangham, has been 'increasingly turning our food to mush, to the maximum calories you can get out of it. Which, of course, is all very ironic, because in the West there's tremendous pressure to reduce the number of calories you're getting out of your food.' He expects to find examples of structural differences that affect caloric availability in many more foods. 'I think there is work here for hundreds and probably thousands of nutritionists for years,' he says.

There's also the problem that no two people are identical. Differences in height, body fat, liver size, levels of the stress hormone cortisol, and other factors influence the energy required to maintain the body's basic functions. Between two people of the same sex, weight and age, this number may differ by up to 600 calories a day – over a quarter of the recommended intake for a moderately active woman. Even something as seemingly insignificant as the time at which we eat may affect how we process energy. In one recent study, researchers found that mice fed a high-fat diet between 9am and 5pm gained 28 per cent less weight than mice fed the exact same food across a 24-hour period. The researchers suggested that irregular feedings affect the circadian cycle of the liver and the way it metabolises food, thus influencing overall energy balance. Such differences would not emerge under the feeding schedules in the Beltsville experiments.

Until recently, the idea that genetics plays a significant role in obesity had some traction: researchers hypothesised that evolutionary pressures may have favoured genes that predisposed some people to hold on to more calories in the form of added fat. Today, however, most scientists believe we can't blame DNA for making us overweight. 'The prevalence of obesity started to rise quite sharply in the 1980s,' says Nestle.

'Genetics did not change in that ten- or twenty-year period. So genetics can only account for part of it.'

Instead, researchers are beginning to attribute much of the variation to the trillions of tiny creatures that line the coiled tubes inside our midriffs. The microbes in our intestines digest some of the tough or fibrous matter that our stomachs cannot break down, releasing a flow of additional calories in the process. But different species and strains of microbes vary in how effective they are at releasing those extra calories, as well as how generously they share them with their host human.

In 2013, researchers in Jeffrey Gordon's lab at Washington University tracked down pairs of twins of whom one was obese and one lean. He took gut microbes from each, and inserted them into the intestines of microbe-free mice. Mice that got microbes from an obese twin gained weight; the others remained lean, despite eating the exact same diet. 'That was really striking,' said Peter Turnbaugh, who used to work with Gordon and now heads his own lab at the University of California, San Francisco. 'It suggested for the first time that these microbes might actually be contributing to the energy that we gain from our diet.'

The diversity of microbes that each of us hosts is as individual as a fingerprint and yet easily transformed by diet and our environment. And though it is poorly understood, new findings about how our gut microbes affect our overall energy balance are emerging almost daily. For example, it seems that medications that are known to cause weight gain might be doing so by modifying the populations of microbes in our gut. In November 2015, researchers showed that risperidone, an antipsychotic drug, altered the gut microbes of mice who received it. The microbial changes slowed the animals' resting metabolisms, causing them to increase their body mass by 16 per cent in two months. The authors liken the effects to a 30-lb weight gain over one year for an average human, which they say would be the equivalent of an extra cheeseburger every day.

Other evidence suggests that gut microbes might affect weight gain in humans as they do in lab animals. Take the case of the woman who gained more than 40 lbs after receiving a transplant of gut microbes from her overweight teenage daughter. The transplant successfully treated the mother's intestinal infection of Clostridium difficile, which had resisted antibiotics. But, as of the study's publication last year, she hadn't been able to shed the excess weight through diet or exercise. The only aspect of her physiology that had changed was her gut microbes.

All of these factors introduce a disturbingly large margin of error for an individual who is trying, like Nash, Haelle and millions of others, to count calories. The discrepancies between the number on the label and the calories that are actually available in our food, combined with individual variations in how we metabolise that food, can add up to much more than the 200 calories a day that nutritionists often advise cutting in order to lose weight. Nash and Haelle can do everything right and still not lose weight.

None of this means that the calorie is a useless concept. Inaccurate as they are, calorie counts remain a helpful guide to relative energy values: standing burns more calories than sitting; cookies contain more calories than spinach. But the calorie is broken in many ways, and there's a strong case to be made for moving our food accounting system away from that one particular number. It's time to take a more holistic look at what we eat.

§ § §

Wilbur Atwater worked in a world with different problems. At the beginning of the 20th century, nutritionists wanted to ensure people were well fed. The calorie was a useful way to quantify a person's needs. Today, excess weight affects more people than hunger; 1.9 billion adults around the world are considered overweight, 600 million of them obese. Obesity brings with it a higher risk of diabetes, heart disease and cancer. This is a new challenge, and it is likely to require a new metric.

One option is to focus on something other than energy intake. Like satiety, for instance. Picture a 300-calorie slice of cheesecake: it is going to be small. 'So you're going to feel very dissatisfied with that meal,' says Susan Roberts. If you eat 300 calories of a chicken salad instead, with nuts, olive oil and roasted vegetables, 'you've got a lot of different nutrients that are hitting all the signals quite nicely,' she says. 'So you're going to feel full after you've eaten it. That fullness is going to last for several hours.'

As a result of her research, Roberts has created a weight-loss plan that focuses on satiety rather than a straight calorie count. The idea is that foods that help people feel satisfied and full for longer should prevent them from overeating at lunch or searching for a snack soon after cleaning the table. Whole apples, white fish and Greek yoghurt are on her list of the best foods for keeping hunger at bay.

There's evidence to back up this idea: in one study, Roberts and colleagues found that people lost three times more weight by following her satiety plan compared with a traditional calorie-based one – and kept it off. Harvard nutritionist David Ludwig, who also proposes evaluating food on the basis of satiety instead of calories, has shown that teens given instant oats for breakfast consumed 650 more calories at lunch than their peers who were given the same number of breakfast calories in the form of a more satisfying omelette and fruit. Meanwhile, Adam Drewnowski, a epidemiologist at the University of Washington, has his own calorie upgrade: a nutrient density score. This system ranks food in terms of nutrition per calorie, rather than simply overall caloric value. Dark green vegetables and legumes score highly. Though the details of their approaches differ, all three agree: changing how we measure our food can transform our relationship with it for the better.

Individual consumers could start using these ideas now. But persuading the food industry and its watchdogs, such as the FDA, to adopt an entirely new labelling system based on one of these alternative measures is much more of a challenge. Consumers are unlikely to see the calorie replaced by Roberts's or Drewnowski's units on their labels any time soon; nonetheless, this work is an important reminder that there are other ways to measure food, ones that might be more useful for both weight loss and overall health.

Down the line, another approach might eventually prove even more useful: personalised nutrition. Since 2005, David

Wishart of the University of Alberta has been cataloguing the hundreds of thousands of chemical compounds in our bodies, which make up what's known as the human metabolome. There are now 42,000 chemicals on his list, and many of them help digest the food we eat. His food metabolome database is a more recent effort: it contains about 30,000 chemicals derived directly from food. Wishart estimates that both databases may end up listing more than a million compounds. 'Humans eat an incredible variety of foods,' he says. 'Then those are all transformed by our body. And they're turned into all kinds of other compounds.' We have no idea what they all are, he adds – or what they do.

According to Wishart, these chemicals and their interactions affect energy balance. He points to research demonstrating that high-fructose corn syrup and other forms of added fructose (as opposed to fructose found in fruit) can trigger the creation of compounds that lead us to form an excess of fat cells, unrelated to additional calorie consumption. 'If we cut back on some of these things,' he says, 'it seems to revert our body back to more appropriate, arguably less efficient metabolism, so that we aren't accumulating fat cells in our body.'

It increasingly seems that there are significant variations in the way each one of us metabolises food, based on the tens of thousands – perhaps millions – of chemicals that make up each of our metabolomes. This, in combination with the individuality of each person's gut microbiome, could lead to the development of personalised dietary recommendations. Wishart imagines a future where you could hold up your smartphone, snap a picture of a dish, and receive a verdict on how that food will affect you as well as how many calories you'll extract from it. Your partner might receive completely different information from the same dish.

Or maybe the focus will shift to tweaking your microbial community: if you're trying to lose weight, perhaps you will curate your gut microbiome so as to extract fewer calories without harming your overall health. Peter Turnbaugh cautions that the science is not yet able to recommend a particular set of microbes, let alone how best to get them inside your gut, but he takes comfort from the fact that our microbial populations are 'very plastic and very malleable' – we already know that they change when we take antibiotics, when we travel and when we eat different foods. 'If we're able to figure this out,' he says, 'there is the chance that someday you might be able to tailor your microbiome' to get the outcomes you want.

None of these alternatives is ready to replace the calorie tomorrow. Yet the need for a new system of food accounting is clear. Just ask Haelle. 'I'm kind of pissed at the scientific community for not coming up with something better for us,' she confesses, recalling a recent meltdown at TGI Friday's as she navigated a confusing datasheet to find a low-calorie dish she could eat. There should be a better metric for people like her and Nash – people who know the health risks that come with being overweight and work hard to counter them. And it's likely there will be. Science has already shown that the calorie is broken. Now it has to find a replacement.

This story was first published on 26 January 2016 by Wellcome on mosaicscience.com

Could 3D printers make body parts?

■ Ian Birrell

John Nhial was barely a teenager when he was grabbed by a Sudanese guerrilla army and forced to become a child soldier. He was made to endure weeks of walking with so little food and water that some of his fellow captives died. Four more were killed one night in a wild-animal attack. Then the boys were given military training that involved 'running up to ten kilometres in the heat and hiding' before being given guns and sent to fight 'the Arabs'.

He spent four years fighting, bombed from the skies and blasting away on guns almost too heavy to hold against an enemy sometimes less than a kilometre away. 'I think, 'If I killed that one it's a human being like me,' but you are forced,' he said. One day the inevitable happened: Nhial (not his real name) was injured, treading on a mine while on early-morning patrol with two other soldiers in a patch of Upper Nile state surrounded by their enemies.

'I stepped on it and it exploded,' he recalled. 'It threw me up and down again – and then I was looking around for my foot. I tried to look for my leg and found that there was no foot. When I saw there's no foot I feel shock. I was really confused. If I was not with the two others I would kill myself because I thought there was no use for me now, so I decide to die.'

His comrades carried him back to base camp, but there was hardly any medical care there. It took 25 days before he received proper treatment, during which time he developed tetanus on one side of his body. Finally Nhial was put on a flight to the Kenyan border, his life saved when he was handed over to a Red Cross health team. Now, a decade later, he lives in a Juba refugee camp, having suffered further troubles in the whirlwind of conflict that has engulfed the struggling new nation of South Sudan.

During one outbreak of violence he was rounded up with other Nuer – the country's second-largest ethnic group – and taken to an army barracks. His life was only spared when he was dismissed as 'useless' because of his disability. Today he plays wheelchair basketball for his country, although he relies on a prosthetic lower leg to struggle his way round the muddy, sprawling camp that entails long walks to reach the most basic services. It can be difficult to get to training. But at least his hands are free to carry things such as food and water, unlike those on crutches.

Mary Lam (not her real name), 34, who caught polio as a child and today works as a restaurant supervisor in the capital, Juba, explained what it was like growing up reliant on bamboo sticks to haul herself around with a bad leg. She would get up much earlier than her siblings, since it took an hour to get to the classroom and they could rush there much quicker. 'It was hard to go with my exercise book to school unless I tie it on my back like a baby,' she said. And it limited her use of her arms too. 'When two hands are using the bamboos you are not able to do domestic work in the house.'

Stories of lives devastated by conflict or disease are all too common across low-income countries. Lack of an arm or leg can be tough anywhere, but for people in poorer parts of the planet, with so much less support and more rickety infrastructure, it is especially challenging. Some are victims of conflict, others were born with congenital conditions. Many more are injured on roads, the casualty toll soaring in low-income nations even as it plummets in wealthier ones. Every minute, 20 people are seriously injured worldwide in road crashes. In Kenya, half the patients on surgical wards have road injuries.

The World Health Organization (WHO) estimates there are about 30 million people like Nhial and Lam who require prosthetic limbs, braces or other mobility devices. These can be simple to make and inexpensive. As one veteran prosthetist told me, his specialism is among the most instantly gratifying areas of medicine. 'A patient comes in on Monday on crutches that leave them unable to carry anything. By Wednesday they are walking on a new leg and on Friday they leave with their life transformed.'

Yet more than eight in ten of those people needing mobility devices do not have them. They take a lot of work and expertise to produce and fit, and the WHO says there is a shortage of 40,000 trained prosthetists in poorer countries. There is also the time and cost to patients, who may have to travel long distances for treatment that can take five days – to assess need, produce a prosthesis and fit it to the residual limb. The result is that unglamorous items such as braces and artificial limbs are among the most-needed devices to assist lives. Yet, as in so many other areas, technology may be hurtling to the rescue, this time in the shape of 3D printing.

§ § §

Slowly but surely, 3D printing – otherwise known as additive manufacturing – has been revolutionising aspects of medicine since the start of the century, just as it has impacted on so many other industries, from cars to clothing. Perhaps this is not surprising, given that its key benefit is to enable rapid and cost-efficient creation of bespoke products. There are, after all, few commercial products that need to suit a wider variety of shapes and sizes than medical devices made for human beings.

Experts have developed 3D-printed skin for burn victims, airway splints for infants, facial reconstruction parts for cancer patients, orthopaedic implants for pensioners. The fast-developing technology has churned out more than 60 million customised hearing-aid shells and ear moulds, while it is daily producing thousands of dental crowns and bridges from digital scans of teeth, disrupting the traditional wax modelling methods used for centuries. Jaw surgeries and knee replacement operations are also routinely carried out using surgical guides printed on the machines.

So it is unsurprising the technology began to stir interest in the field of prosthetics – even if sometimes by accident.

Ivan Owen is an American artist who likes to make 'weird, nerdy gadgets' for use in puppetry and budget horror movies. In 2011 he created a simple metal mechanical hand for a steampunk convention, the spiky fingers pulled by loops through his own. He posted a video that – as is the way in our interconnected world – was seen by a carpenter in South Africa who had just lost four fingers in a circular saw accident. They began discussing plans for a prototype prosthetic hand, and soon that came to the attention of the mother of a five-year-old boy called Liam, born without fingers on his right hand.

She wanted a tiny version of their hand. But Owen realised the child would rapidly grow out of anything they made, so he looked at the idea of using 3D printing. 'If we could develop a design that was printable, it would be possible to rescale and reprint that design as Liam grew, essentially making it possible for his device to grow with him,' he said. So the artist persuaded a printer manufacturer to donate two machines and developed what has been claimed to be the first 3D-printed mechanical hand. And crucially, rather than patent this work, Owen published the files as open source for anybody to access, allowing others to collaborate, to use and improve the designs.

This has grown into Enabling the Future, a community with 7,000 members in dozens of countries and access to 2,000 printers, who help make arms and hands for those in need. One school student in California even printed a new hand for a local teacher. Often they are aimed at children, since many dislike the weight, look and hassle of modern prosthetics, which can involve inserting the arm in a silicone sleeve and using straps across the back to hold the device in place. These body-powered hands cost thousands of pounds, yet must be replaced every couple of years as a child grows. The 3D-printed versions cost about £40, come in any colour and look like a cheery toy, so are often more appealing despite being less sophisticated.

Jorge Zuniga, a research scientist in the Biomechanics Research Building at the University of Nebraska in Omaha, heard about this project on his car radio. He was only half-listening, but arriving home he started playing baseball with his four-year-old son and observed how important the grabbing of an object was to his own child's development. He spent the next month carefully building a prosthetic model that mimicked the human hand, only for his work to be dismissed instantly by his son. 'He told me children wanted a hand that looked like a robot.'

From this conversation and the open-source designs available emerged Cyborg Beast, a project being heavily backed by his department to develop futuristic-looking, low-cost prosthetic hands. 'You can do anything with 3D printing,' said Zuniga, who now heads a seven-strong team. 'We believe it will revolutionise the prosthetics field. It will lower the costs worldwide and gives engineers, patients and doctors the chance to modify prosthetic hands as they want. And they can be any colour.'

When I told Zuniga slightly hesitantly that his design looked like a toy, he was delighted. 'That's great – we want children to see it as a toy,' he said. 'This is a transitional device. Many children do not like prosthetics, however good they are these days, because they might have a hook for a hand and the harness needs help to put on, which children dislike. So this is to bridge the gap, helping them get used to the idea as they grow up.

'We have even had a child missing a shoulder. So we developed a device that weighs the same as the missing arm. This meant he not only got a new arm that helped daily life but it also improved his posture and balance, therefore was much better for his spine. This sort of thing can be done much easier with 3D technology. But of course the difference between a toy and a prosthetic arm is that you need professional involvement to enhance use of the devices and ensure they are fitting properly.'

It is remarkable that people who do not even own a printer can obtain a functional child's hand for the price of a theatre ticket within 24 hours. Zuniga says at least 500 Cyborg Beasts are in use worldwide, and the design has been downloaded almost 50,000 times. He has taken it to his native Chile, where he runs a paediatric orthopaedic 3D-printing laboratory, and has had recent requests for the plans from Nigeria. 'My concern at this stage is that some of the materials can melt in higher temperatures. It is not working well there yet, but this sort of prosthetic has huge potential to be used with better materials in the developing world. We are still in the infancy stage at this moment.'

Another place that has experimented with this technology is in the cruel, forgotten war cursing the Nuba mountains of Sudan, where an amazing American named Tom Catena has been working as the only permanent doctor for half a million people around his Mother of Mercy Hospital. Fuelled by his religious faith, for almost a decade this brave medic has ignored bombings, lack of electricity and water shortages to do everything from delivering babies to amputating limbs.

'It's demoralising for us to amputate an arm knowing that there is no good solution,' Catena told me by email. 'We have many arm amputees – both above and below the elbow as a result of the war here and general lack of medical care. This in an agricultural society, where nearly everyone is a subsistence farmer. If one is missing an arm, he is not very functional in this society. They become totally dependent on the family and they have a difficult time getting married (also very important in this society).'

The idea of using 3D printing to help arose when Mick Ebeling, an American film producer and philanthropist, learned about this work at the same time as he was hearing about the emerging work on low-cost prosthetic hands. Searching for information on Catena, Ebeling read about one of his patients: Daniel Omar, a 12-year-old boy who had wrapped his arms around a tree to protect himself during an aerial attack. His face and body were protected when a bomb exploded nearby – but both the boy's arms were blown off.

Ebeling travelled out with printers and, working with hospital staff, fitted about a dozen people with new arms. 'Unfortunately, as time went on, none of the amputees were using the prostheses as they felt they were too cumbersome,' said Catena. The doctor concluded that 'the 3D model was good, fairly easy to make and inexpensive... although it hasn't worked out so well here, perhaps with some tweaking, the 3D printers can be of great use for arm amputees.'

§ § §

Yet for all the agonies and difficulties associated with arm loss, the bigger problem in low-income countries is when lower limb disability leads to loss of mobility. Wheelchairs are expensive and can be difficult to use when roads are pot-holed, streets are muddy and pavements are non-existent. Without a prosthetic limb, people struggle to fetch water, to prepare food and, above all, to work. This throws them back on their families and communities, intensifying any hardship and poverty.

One group that has spent almost three decades trying to tackle such issues is Exceed, a British charity set up by diplomats and academics at the request of Cambodia's government to help thousands of landmine survivors. It works in five Asian countries, training people at schools of prosthetics and orthotics. In Cambodia, there are still almost 9,000 landmine survivors in need of artificial limbs, although these days traffic accidents are a more likely cause of disability, while children also need braces for a range of common conditions such as spina bifida, cerebral palsy and polio.

'If you wear a prosthesis you are disabled for about ten minutes in the morning while you have a shower, then you put your leg on and go to work. If you do not have one, then your hands are out of use with crutches so you can't even take drinks to the table,' said Carson Harte, a 59-year-old prosthetist and chief executive of Exceed. 'Without a prosthesis there are no expectations. You just go back and rely on the goodwill of your family.'

It is not really cash shortages that deny people these devices, since simplified forms cost little and generic Chinese models are improving fast. The components can cost just £30. The big hurdle is the lack of trained technicians to fit the artificial limbs. In the Philippines, there are estimated to be 2 million people needing prosthetics or orthotics. Yet there are only nine fully trained experts, each able to assist at most 400 patients a year with the time-consuming process of creating and fitting a customised limb, although more are being trained on a new four-year course.

Traditionally, a prosthetist would wrap a stump with plaster of Paris bandages to make a reverse mould and let it dry, then fill it with more plaster that must harden. From this a socket can be forged that fits, with more modifications for precision, to the bone on the stump. Great care must be taken to avoid nerves and tender areas that are not tolerant of pressure. The key for the technician is to understand the pathology of a stump, which differs for each person. This is a cumbersome process that can take a week, especially with gait training for new patients that lasts three days. It can also be messy work, mixing up and moulding the plaster, while a prosthetist visiting a rural area must cart around 20-kilo packs of plaster. But with a 3D scanner, a digital image can be made in half an hour and sent by email, and there is no mess.

Exceed has begun a seven-month trial of 3D-printed devices in Cambodia with Nia Technologies, an innovative Canadian not-for-profit organisation. 'This technology has the potential to increase the productivity of every technician,' said Harte. 'It is not about printing off legs, nor does it replace the skills of a well-trained professional, but it has potential to produce a better, faster, more easily repeatable way of doing one key part of the chain. There are no magic bullets, but this may be an important incremental change.

'The key to success so far has been cross-fertilisation: putting engineers and prosthetist orthotists together. Engineers make broad assumptions that are not always right, prosthetist orthotists do not always know what engineers can do. Together we have made more advances in a few months than have been achieved in years, sorting our real problems in real time through collaboration.'

Nia is also trialling its 3D PrintAbility technology in Tanzania and Uganda, where there are only 12 prosthetists to serve a population of about 40 million people – and at the time of writing all six state clinics have run out of materials. Doctors there often deal with children who have lost limbs after falling in open cooking fires, while other youngsters need braces after suffering post-injection paralysis caused by badly administered jabs that damage nerves.

In Uganda their team is working with CoRSU hospital in Kisubi, a specialist rehabilitation centre for children with disabilities. Orthopaedic technician Moses Kaweesa said they found the technology lighter and faster to use, as well as easier for people in remote rural areas. 'It used to take five days to have a limb manufactured, with lots of hanging around for the patient. Now it is barely two days, so they spend much less time in the hospital. There is also less waste of material, so for a country like ours this can help so much by cutting down the costs.'

The first person to test out a 3D-printed mobility device was a four-year-old girl who until then had dragged herself across floors and had to be carried around by her family. 'When she was born her leg was missing the right foot,' said her older brother. 'It was very difficult for her to walk, to play with other children. She can be lonely. But when she was given a leg she was able to run with others, play with others.'

Matt Ratto, Nia's chief science officer, who led the project's development, admitted that it was only when he saw the serious-looking child in her red dress start to walk that he realised his technology actually worked. But, like Harte, he urges caution. 'We are surrounded by the hype of 3D printing with crazy, ridiculous claims being made,' he said. 'We must be cautious. A lot of these technologies fail not for engineering reasons but because they are not designed for the developing world. You can't just smash in these new technologies.

'A lot of what we are doing is social innovation. People think you are threatening to replace prosthetists, which is a problem since they can be hesitant to embrace it just like in the developing world. We are trying to develop a bridge between the North and South but we have to work with the people on the ground to build their capacity. They are the experts – and they are deeply interested in doing whatever they can to get the children walking.'

Ratto's aim is to use their technology to fit 8,000 people with 3D-printed mobility devices within five years, across some 20 sites in low-income countries. 'My sense if we get this right is that the growth can be exponential. If we iron out the kinks and work out the best way to help clinicians I think we will see something of a hockey-stick curve on the graph. But we must not get it wrong, move too fast nor over-hype the potential.'

§ § §

One person who shares Ratto's belief in this technology is Claudine Humure, a 24-year-old Rwandan with big ambitions whom I met on a chilly November day at Wheaton College in Norton, Massachusetts. She lost her parents in the genocide, and then at the age of 12, while living in an orphanage, developed a dreadful pain in her right leg that turned out to be bone cancer. 'I thought I was going to die when the doctors told me, because that was all I knew about cancer. I thought that was my fate,' she said. 'Then they said they had to amputate my leg the next day. I was crying so hard. I hated the doctors who were telling me the news since everything was happening so fast.'

After the operation, Humure was flown by a US charity to Boston for further treatment, including chemotherapy and surgery to prepare the leg for a prosthetic. She spent almost a year recovering before returning to Rwanda with an artificial leg. But when it broke, she struggled to find a new one and saw the deficiencies of what was available in low-income countries. 'I had seen what was possible. A good prosthesis fits well and feels comfortable. You can do anything with it, you feel normal.'

Now this affable woman is back in the US, studying biology and business at a prestigious university. She wrote her high-school project on the design of prosthetic limbs and has volunteered at Spaulding Rehabilitation Hospital in Boston, where she spent time with injured victims of the 2013 marathon bombing. 'They were very new to the idea of missing a limb, so it was very traumatic for them. You could see they were terrified since it was so sudden. I hope I was a positive influence, an inspiration not to give up hope.'

Most significantly, Humure won a biomechatronics research internship at Massachusetts Institute of Technology's Media Lab. Here she met Hugh Herr, a pioneering figure in prosthetics. He lost both legs after being trapped for three nights in a freezing blizzard while climbing, then designed titanium-tipped artificial feet to let him return to his beloved mountains. She also came across 3D printing for the first time. 'This was life-changing,' she said. 'My eyes opened. I saw all this cutting-edge research when we had such bad prostheses in Rwanda. I looked at my own prosthetic leg and started thinking.'

Now she dreams of opening specialist clinics, first in Rwanda, then across the rest of Africa. And she is designing a socket for prosthetic limbs to be used by people who have had leg amputations above the knee, aimed at low-income nations. 'I am making the socket lighter, easier to use and cheaper to manufacture. But what makes the design special is that the user can adjust it to make it more comfortable. In developing countries, people just do not have the time to keep travelling to clinics.'

Humure believes such advances can change the world for millions of people like her. 'You can have a disability and still be successful. I know I have been lucky in many ways because I met the right people, but I am a positive person and this is the attitude I want to instil in other amputees and people with disabilities, especially the millions of us in developing countries. A good prosthesis does not just help your mobility. It gives you confidence and can change your life. Above all, you forget you are an amputee.'

This story was first published on 19 February 2017
by Wellcome on mosaicscience.com

How to fall to your death and live to tell the tale

■ Neil Steinberg

Alcides Moreno and his brother Edgar were window washers in New York City. The two Ecuadorian immigrants worked for City Wide Window Cleaning, suspended high above the congested streets, dragging wet squeegees across the acres of glass that make up the skyline of Manhattan.

On 7 December 2007, the brothers took an elevator to the roof of Solow Tower, a 47-storey apartment building on the Upper East Side. They stepped onto the 16-foot-long, three-foot-wide aluminium scaffolding designed to slowly lower them down the black glass of the building.

But the anchors holding the 1,250-pound platform instead gave way, plunging it and them 472 feet to the alley below. The fall lasted six seconds.

Edgar, at 30 the younger brother, tumbled off the scaffolding, hit the top of a wooden fence and was killed instantly.

Part of his body was later discovered under the tangle of crushed aluminium in the alley next to the building.

But rescuers found Alcides alive, sitting up amid the wreckage, breathing and conscious when paramedics performed a 'scoop and run' – a tactic used when a hospital is near and injuries so severe that any field treatment isn't worth the time required to do it. Alcides was rushed to NewYork-Presbyterian Hospital/Weill Cornell Medical Center, four blocks away.

§ § §

Falls are one of life's great overlooked perils. We fear terror attacks, shark bites, Ebola outbreaks and other minutely remote dangers, yet over 420,000 people die worldwide each year after falling. Falls are the second leading cause of death by injury, after car accidents. In the United States, falls cause 32,000 fatalities a year (more than four times the number caused by drowning or fires combined). Nearly three times as many people die in the US after falling as are murdered by firearms.

Falls are even more significant as a cause of injury. More patients go to emergency rooms in the US after falling than from any other form of mishap, according to the Centers for Disease Control and Prevention (CDC), nearly triple the number injured by car accidents. The cost is enormous. As well as taking up more than a third of ER budgets, fall-related injuries often lead to expensive personal injury claims. In one case in an Irish supermarket, a woman was awarded 1.4 million euros compensation when she slipped on grapes inside the store.

It makes sense that falls dwarf most other hazards. To be shot or get in a car accident, you first need to be in the vicinity of a gun or a car. But falls can happen anywhere at any time to anyone.

Spectacular falls from great heights outdoors like the plunge of the Moreno brothers are extremely rare. The most dangerous spots for falls are not rooftops or cliffs, but the low-level, interior settings of everyday life: shower stalls, supermarket aisles and stairways. Despite illusions otherwise, we have become an overwhelmingly indoor species: Americans spend less than 7 per cent of the day outside but 87 per cent inside buildings (the other 6 per cent is spent sitting in cars and other vehicles). Any fall, even a tumble out of bed, can change life profoundly, taking someone from robust health to grave disability in less than one second.

Falling can cause bone fractures and, occasionally, injuries to internal organs, the brain and spinal cord. 'Anybody can fall,' says Elliot J Roth, medical director of the patient recovery unit at the Shirley Ryan AbilityLab in Chicago. 'And most of the traumatic brain injury patients and spinal cord injury patients we see had no previous disability.'

There is no *Journal of Falls*, though research into falling, gait and balance has increased tremendously over the past two decades. Advances in technology improve our understanding of how and why people fall, offer possibilities to mitigate the severity of falls, and improve medicine's ability to treat those who have hurt themselves falling.

Scientists are now encouraging people to learn how to fall to minimise injury – to view falling not so much as an unexpected hazard to be avoided as an inevitability to be prepared for. Training may even have been a factor determining the outcome of the Moreno brothers' fall to earth nearly ten years ago.

§§§

Doctors at NewYork-Presbyterian did not want to risk moving Alcides Moreno from the emergency room into a surgical theatre for fear that the slightest additional bump might kill him. They started surgery in the ER. He had two broken legs, a broken arm, a broken foot, several broken ribs, and a crushed vertebra that could have paralysed him, as well as two collapsed lungs, a swollen brain, plus several other ruptured organs. Alcides was given 24 pints of blood and 19 pints of plasma before the bleeding could be stopped.

Doctors marvelled that he was alive at all, reaching for an explanation not often used in medical literature: 'miracle'.

By 100 feet or more, falls are almost always fatal, apart from freak accidents. People have fallen miles from planes and lived, due to tumbling down snowy hillsides, the way extreme skier Devin Stratton did when he accidentally skied off a 150-foot Utah mountain cliff in January 2017 and escaped unharmed, his fall arrested by branches and cushioned by deep snow. He was wearing a helmet, which cracked even as its camera recorded his plunge.

'It's not the fall that gets you,' the skydiving joke goes. 'It's the sudden stop at the bottom.' Deceleration is the key to surviving falls and reducing injuries – it isn't the length of fall that's relevant, but what happens as you reach the ground. This was dramatically demonstrated in the summer of 2016 by professional skydiver and safety expert Luke Aikens. He jumped from a plane without a parachute at an altitude of 25,000 feet, or 4.7 miles, hitting a 100-by-100-foot net positioned in the southern California desert and emerging without a scratch.

One theory was that Alcides lived because, when the scaffolding gave way, he lay flat and clung to the platform, as professional window washers are trained to do. The scaffold fell not in the open street but in a narrow alley – air resistance may have built up against the platform, slowing it. The platform also may have scraped against the building and its neighbour, reducing its rate of fall. The aluminium crushed on impact, and landed on a pile of cables, both of which absorbed some of the impact, forming a cushioned barrier.

Survival from heights prompted the first medical writing about falls. Hippocrates, in his treatise on head injuries, observes, sensibly, that 'he who falls from a very high place upon a very hard and blunt object is in most danger of sustaining a fracture... whereas he that falls upon more level ground, and upon a softer object, is likely to suffer less injury'. The first modern medical paper on a fall was Philip Turner's 'A fall from a cliff 320 feet high without fatal injuries', published in the Guy's Hospital Gazette in 1919. It examined the case of a Canadian Army private who stumbled over a chalk cliff on the coast of France in 1916 and lived.

In 1917, an American air cadet named Hugh DeHaven was flying in a Curtiss JN-4 'Jenny' when it collided with another biplane 700 feet above an airfield in Texas. Among the four men aboard the two planes, DeHaven alone survived the plunge. He spent the rest of his career trying to figure out why, culminating in his pioneering 1942 paper, 'Mechanical analysis of survival in falls from heights of fifty to one hundred and fifty feet'.

In it, he examined eight cases of people surviving long falls – ignoring his own, but including the lucky Canadian private from 1916 – and found that those who landed on newly tilled gardens could walk away surprisingly intact, noting: 'It is, of course, obvious that speed or height of fall, is not in itself injurious.' That might sound like the first half of the skydiving joke, but his research led him to design and patent the combined seatbelt and shoulder harness worn in every car today.

Up to the 1960s and 1970s, scientific papers on falls focused on forensics – their subjects tended to be dead, the medical questions centring on what had happened to them. This was important, for instance, when assessing trauma to children – could this child have fallen and suffered these injuries, as the caregiver claimed, or is it abuse? Falls as a separate, chronic, survivable medical problem began to get attention only in the past quarter-century. The journal Movement Disorders was begun in 1986, but the bulk of papers examining the interplay of balance, gait and falls at ground level appear after 2000.

§ § §

You can trip or slip when walking, but someone standing stock still can fall too – because of a loss of consciousness, vertigo or, as the Moreno brothers remind us, something supposedly solid giving way. However it happens, gravity takes hold and a brief, violent drama begins. And like any drama, every fall has a beginning, middle and end.

'We can think of falls as having three stages: initiation, descent and impact,' says Stephen Robinovitch, a professor in the School of Engineering Science and the Department of Biomedical Physiology and Kinesiology at Simon Fraser University in British Columbia, Canada. 'Most research in the area of falls relates to 'balance maintenance' – how we perform activities such as standing, walking and transferring without losing balance.'

By 'transferring', he means changing from one state to another: from walking to stopping, from lying in a bed to standing, or from standing to sitting in a chair. 'We have found that falls among older adults in long-term care are just as likely to occur during standing and transferring as during walking,' says Robinovitch, who installed cameras in a pair of Canadian nursing homes and closely analysed 227 falls over three years.

Only 3 per cent were due to slips and 21 per cent due to trips, compared to 41 per cent caused by incorrect weight shifting – excessive sway during standing, or missteps during walking. For instance, an elderly woman with a walker turns her upper body and it moves forward while her feet remain planted. She topples over, due to 'freezing', a common symptom of Parkinson's, experienced regularly by about half of those with the disease.

In general, elderly people are particularly prone to falls because they are more likely to have illnesses that affect their cognition, coordination, agility and strength. 'Almost anything that goes wrong with your brain or your muscles or joints is going to affect your balance,' says Fay Horak, professor of neurology at Oregon Health & Science University.

Fall injuries are the leading cause of death by injury in people over 60, says Horak. Every year, about 30 per cent of those 65 and older living in senior residences have a fall, and when they get older than 80, that number rises to 50 per cent. A third of those falls lead to injury, according to the CDC, with 5 per cent resulting in serious injury. It gets expensive. In 2012, the average hospitalisation cost after a fall was $34,000.

How you prepare for the possibility of falling, what you do when falling, what you hit after falling – all determine whether and how severely you are hurt. And what condition you are in is key. A Yale School of Medicine study of 754 over-70s, published in the Journal of the American Medical Association in 2013, found that the more serious a disability you have beforehand, the more likely you will be severely hurt by a fall. Even what you eat is a factor: a study of 6,000 elderly French people in 2015 found a connection between poor nutrition, falling and being hurt in falls.

§ § §

Alcides Moreno underwent 15 more surgeries and was in a coma for weeks. He was visited by his three children: Michael, 14, Moriah, 8, and Andrew, 6. His wife, Rosario, stayed at his bedside, talking to him. She repeatedly took his hand and guided it to stroke her face and hair, hoping that the touch of her skin would help bring him around. Then, on Christmas Day, Alcides reached out and stroked not his wife's face but the face of one of his nurses.

'You're not supposed to do that,' Rosario chided him. 'I'm your wife. You touch your wife.'

'What did I do?' he asked. It was the first time he had spoken since the accident, 18 days earlier. His doctors predicted he might walk again, after lengthy rehabilitation, though the challenges proved to be not only physical but also mental. People who fall suffer the expected physical injuries, but accidental falling also carries a heavy psychological burden that can make recovery more difficult and can, counter-intuitively, set the stage for future falls.

§ § §

Children begin to walk, with help, at about a year old. By 14 months they are typically walking unaided. Those first baby steps are guided by three key bodily systems. First, proprioception – input from the nerves in the muscles, a sense of where limbs are relative to each other and what they're doing. People whose limbs are numb have difficulty walking even if their musculature is completely functional.

The second sense is vision, not just to see where you are going, but to help process information from your other senses. 'Most people who walk into a dark room will sway more than if they can see – about 20 per cent more,' says Horak, an expert in how neurological disorders affect balance and gait.

And third is your vestibular system, canals of fluid in the inner ear that work in a way not very different from a carpenter's spirit level. The system takes measurements in three dimensions, and your body uses the data to orient itself.

With these various systems doing their jobs, you can step forward and begin to walk, a feat that performance artist Laurie Anderson once described in a song with succinct scientific accuracy: 'You're walking, and you don't always realise it/But you're always falling/With each step, you fall forward slightly/And then catch yourself from falling/Over and over.'

Or don't catch yourself. We fall when the smooth, almost automatic process of walking goes awry. Perhaps it is something as crude as your step being blocked by an obstacle: you trip, over a prankster's outstretched foot perhaps. Or the traction of your foot against the floor is lost because of a slippery substance – the classic banana peel of silent movie fame, what researchers call 'perturbation'.

Christine Bowers is 18. She hails from upstate New York, and is a student at the Moody Bible Institute in Chicago. One day she hopes to teach English abroad. In January 2016 she had a cavernous malformation – a tangle of blood vessels deep within her brain – removed.

'It paralysed my left side,' she says, as her physical therapist straps her into a complex harness in a large room filled with equipment at the Shirley Ryan AbilityLab. 'I'm working on preventing a fall.'

Under the supervision of Ashley Bobich the therapist, Bowers is walking on the KineAssist MX, a computerised treadmill with a robotic arm and harness device at the back. The metal arm allows patients freedom of motion but catches them if they fall. This version of the device is quite new – the AbilityLab only got it at the end of 2016 and Bowers is the second patient of Bobich's to try it. Previously, those in danger of falling would be tethered to overhead gate tracks, a far cruder system, which still can be seen in the ceilings above.

Being a student, Bowers often finds herself in crowded academic hallways, and says she values her cane as much to alert those around her that she has mobility problems as for support. Seeing the cane, she says, her classmates tend to give her a bit of room as they hurry through the corridors.

Still, she has fallen several times, and those falls made her very skittish about walking, a serious problem in the rehabilitation of those who have fallen. 'It's huge,' says Bobich. 'Fear of falling puts you at risk for falling.'

Elliot Roth agrees. 'Falls often cause fear of falling, and fear of falling often causes fear of walking, and fear of walking often causes abnormal or inadequate walking,' he says. A challenge of rehabilitation is to not only increase physical capacity, but also build patient confidence.

'We've been doing what's called 'perturbation training', where I pick a change in the treadmill speed,' says Bobich. 'She's walking along, I hit the button, and the treadmill speeds up on her and she has to react... Her biggest fear was slipping on ice, so I said, 'You know what? I have a really great way for us to train that.''

The treadmill hums while Bobich speeds it up and slows it down, and Bowers, her right hand clasping her paralysed left, struggles to maintain her balance.

'You're getting better at this,' says Bobich. 'You're getting way better.'

§ § §

The KineAssist is an example of how technology that was once used to study ailments is now used to help patients. Advanced brain scanning, having identified the regions responsible for balance, now diagnoses damage that affects them. Accelerometers attached to people's ankles and wrists have been used in experiments, plotting induced falls directly into a computer for study, and are now being used to diagnose balance problems – or to detect when someone living alone has fallen and summon help.

'Over one-half of older adults who fall are unable to rise independently, and are at risk for a 'long lie' after a fall, especially if they live alone, which can greatly increase the clinical consequences of the falls,' says Stephen Robinovitch. He and his colleagues are working to develop wearable sensor systems that detect falls with high accuracy, as well as providing information on their causes, and on near-falls.

Researchers at the Massachusetts Institute of Technology took the 'wearable' out of the equation by developing a radio wave system that detects when someone has fallen and automatically summons help. The Emerald system was shown off at the White House in 2015 but is still finding its way to a market chock-full of devices that detect falls, invariably pendants.

Not that a device needs to be high-tech to mitigate falls. Wrestlers use mats because they expect to fall; American football running backs wear pads. Given that a person over 70 is three times as likely to fall as someone younger, why don't elderly people generally use either?

The potential benefit of cushioning is certainly there. The CDC estimates that $31 billion a year is spent on medical care for over-65s injured in falls – $10 billion for hip fractures alone (90 per cent of which are due to falls). Studies show that such pads reduce the harmful effects of falling.

But older people have all the vanity, inhibition, forgetfulness, wishful thinking and lack of caution that younger people have, and won't wear pads. More are carrying canes and using walkers than before, but many more who could benefit shun them because, to them, canes and walkers imply infirmity, a fate worse than death (80 per cent of elderly women told researchers in one study that they would rather die than have to live with a debilitating hip fracture). This sets up another vicious cycle related to falling: fearing the appearance of disability, some elderly people refuse to use canes, thereby increasing their chances of falling and becoming disabled.

Padded floors would seem ideal, since they require none of the diligence of body pads or canes. But padding environments is both expensive and a technical challenge. If a flooring material has too much give, wheelchairs can't roll and footing is compromised. That's why nursing homes tend not to be thickly carpeted. People pick up their feet less high as they age, and so have a tendency to trip on carpets.

There are materials designed to reduce injuries from falls. Kradal is a thin honeycombed flooring from New Zealand that transmits the energy of a fall away from whatever strikes it, reducing the force. A study of the flooring in Swedish nursing homes found that while it did reduce the number of injuries when residents fell on it, they fell more frequently when walking on it, leading to a dilemma: the flooring might be causing some falls even as it reduced the severity of resulting injuries.

One unexpected piece of anti-fall technology is the hearing aid. While the inner ear's vestibular system is maintaining balance, sound itself also seems to have a role.

'We definitely found that individuals with hearing loss had more difficulty with balance and gait, and showed significant improvement when they had a hearing aid,' says Linda Thibodeau, a professor at the University of Texas at Dallas's Advanced Hearing Research Center, summarising a recent pilot study. 'Most people don't know about this.'

Horak agrees, saying that people who have cochlear implants to give them hearing also find their balance improves. Hearing is not as critical for balance as proprioception, vision and the vestibular system, she says, 'But hearing may also contribute and we don't understand how. We think you can use your hearing to orient yourself.'

Thibodeau says one reason it's important to establish this link is that insurance companies don't typically cover hearing aids, because they are seen as improving lifestyle more than sustaining basic health. Hearing aids can be expensive – up to $6,000 – but a broken hip, which insurance companies do cover, can cost five or ten times that figure, or more, and lead to profound disability or death.

More than half of people in their 70s have hearing loss, but typically wait ten to 20 years beyond the time when they could first benefit before they seek treatment. If the connection to balance and falls were better known, that delay might be reduced.

The role of hearing reminds us that, while walking is considered almost automatic, balance is at some level a cognitive act, achieved by processing a cloud of information. Pile demands on our attention and that itself can cause falls, particularly among people who are already compromised physically or cognitively.

Thibodeau once led a group of people with hearing impairments to the Dallas World Aquarium to test out wireless microphone technology in the real world. 'There's a stairway going by an enormous fish tank,' she says. 'I had a participant fall on the stairs, and someone at the aquarium told me, 'A lot of people fall going down those stairs, looking at the aquarium." (Asked to comment whether this indeed is a common problem there, the Dallas World Aquarium director did not reply, a reminder perhaps that the legal aspects of falls can inhibit dissemination of information about them).

§ § §

Given the tremendous cost of falls to individuals and society, and the increasing knowledge of how and why falls occur, what can you do to prevent them? And can you do anything to lessen harm in the split second after you start to fall?

1. Prepare your environment

Secure loose rugs or get rid of them. Make sure the tops and bottoms of stairs are lit. Clean up spills immediately. Install safety bars in showers and put down traction strips, and treat slick surfaces such as smooth marble floors with anti-slip coatings. If there's ice outside your home, clear it and put down salt.

2. Fall-proof your routine

Watch where you are going. Don't walk while reading or using your phone. Always hold handrails – most people using stairways do not. Don't have your hands in your pockets, as this reduces your ability to regain your balance when you stumble. Remember that your balance can be thrown off by a heavy suitcase, backpack or bag.

Roth asks most of his patients who have fallen to describe in detail what happened. 'Sometimes people are not paying attention. Multi-tasking is a myth, and people should try very hard to avoid multi-tasking. No texting while walking.'

The more problems you have controlling your balance, the more attention is required, says Horak. 'If you're carrying a big backpack on a slippery log, you don't want somebody to ask you what's for dinner.'

3. Improve your gear

Wear good shoes with treads. On ice, wear cleats – you can buy inexpensive soles with metal studs that slip over your shoes. Do not wear high heels, or at least have a second pair of flat shoes for walking between locations. Get a hearing aid if you need one. Wear a helmet when bicycling, skiing and skateboarding. Use a cane or walker if required. Hike with a walking stick.

4. Prepare your body

Lower body strength is important for recovering from slips, upper body strength for surviving falls. Martial arts training can help you learn how to fall. Drugs and alcohol are obviously a factor in falls – more than half of adult falls are associated with alcohol use – as is sleep apnoea. Get a balanced diet to support bone density and muscle strength. If you feel light-headed or faint, sit down immediately. Don't worry about the social graces, you can get back up once you've established you are not going to lose consciousness.

Understandably, some elderly people fear falling so much that they don't even want to contemplate it. 'People should know they could improve their balance with practice, even if they have a neurological problem,' says Horak.

5. Fall the right way

What happens once you are falling? Scientists studying falling are developing 'safe landing responses' to help limit the damage from falls. If you are falling, first protect your head – 37 per cent of falls by elderly people in a study by Robinovitch and colleagues involved hitting their heads, particularly during falls forward. Fight trainers and parachute jump coaches encourage people to try not to fall straight forward or backward. The key is to roll, and try to let the fleshy side parts of your body absorb the impact.

'You want to reach back for the floor with your hands,' says Chuck Coyle, fight director at the Lyric Opera of Chicago, describing how he tells actors to fall on stage. 'Distribute the weight on the calf, thigh, into the glutes, rolling on the outside of your leg as opposed to falling straight back.'

Young people break their wrists because they shoot their hands out quickly when falling. Older people break their hips because they don't get their hands out quickly enough. You'd much rather break a wrist than a hip.

§ § §

Alcides Moreno underwent a long regimen of physical and occupational therapy at the Kessler Institute for Rehabilitation in New Jersey, working to strengthen his legs, restore his balance, and walk. Occupational therapy was necessary, as well as counselling, as he had grown depressed over the loss of his brother, Edgar.

He is unable to return to work but received a multimillion-dollar settlement in his lawsuit against the scaffolding company, Tractel, after a Manhattan court found that it had installed the platform negligently. The sum wasn't revealed, but a source said it was more than the $2.5 million that Edgar's family received.

Alcides and his family moved to Arizona, and live outside Phoenix. 'This weather is good for my bones,' he told the New York Post. He keeps busy, driving his kids to school and to sporting events, and likes to work out in the gym.

Last year he and his wife had a fourth child, a son.

'I keep asking myself why I lived,' he told the BBC this year. 'I have a new baby – he must be the reason, to raise this kid and tell him my history.'

This story was first published on 6 June 2017 by Wellcome on mosaicscience.com

Miscarriage: the race to protect pregnancies

■ Holly Cave

'Don't worry, pregnancy isn't an illness,' said my midwife, smiling affectionately as I worried about my lack of morning sickness. She must have been well acquainted with the limbo of early pregnancy, the constant fluttering between hope and fear.

Two days later, doubled over on the toilet and clutching a hot water bottle as I watched dark clots of blood drip into the pan, it felt very much like an illness. I knew something was desperately wrong.

The largest lump of tissue – what I believe to be the yolk sac – was smaller than it felt in my heart. I searched for the embryo inside it until my clothes were stained with blood. I couldn't flush the toilet for an hour because I was sure that my baby was in there. Rationality had ceased to register through the distress.

The list of things I don't understand about my miscarriage seems never-ending. I don't know how old the embryo was when it stopped living. I don't know why it stopped living. I will never know.

'Why?' I asked myself. Again and again and again, as if it was a mantra that would take me back in time and stop it happening. Why? If someone could answer that, then at least I'd be able to grapple with another looming question: Will it happen to me again?

§ § §

'Miscarriages are so common – one in five pregnancies end up in a miscarriage,' says Arri Coomarasamy, a professor of gynaecology at the University of Birmingham. Empathy is soothingly evident in his voice as I come to the end of my story.

The one in five figure is often quoted. Sometimes it creeps up to one in four. This is because it's difficult to determine how many miscarriages take place. In the UK, miscarriage means the loss of a pregnancy during the first 23 weeks (any later and it is called stillbirth). But it often occurs before a woman even realises she's pregnant, and most of the time – 85 per cent – it is in the first 12 weeks of pregnancy.

That has given us clear social guidelines. The '12-week rule' warns against telling anyone you're pregnant until the end of the third month. It anticipates the risk of loss, even sets us up to tentatively expect miscarriage during the early stages of pregnancy, but this silence doesn't make it any easier if it does happen.

A recent survey of over 6,000 women who had had a miscarriage, conducted by the charity Tommy's, found that around two-thirds found it hard to talk about. The same number felt that they couldn't discuss their miscarriage with their best friend. A third didn't feel that they could even talk to the father about it.

Finding support remains a challenge for women experiencing miscarriage. Sharing was important for me – although saddening, I took comfort from the fact that friends of mine had also been through it. Like them, I would get through it. But we never talked about the experience itself, the physical process and the effects of miscarriage. Saying 'I had one, too' seemed to be as far as it went.

So here goes. I was nine weeks pregnant when I started bleeding in the middle of a late night shift at work. The sight of that fresh, bright red blood was a sudden, vicious smack in the face. I pressed my hand over my mouth until I could feel the outline of each tooth, as if to prevent anything else leaving my body. I bled; I cramped; I googled. The lady who answered the phone at the community midwife centre directed me to A&E. Later, my GP assured me that I was right not to go. 'I can't think of a worse place to have a miscarriage,' he said, his head in his hands.

The pain was bearable and the bleeding stopped after a week or so. My miscarriage was natural and complete, meaning that when I had a scan at the end of it, there was barely any evidence that I'd been pregnant at all. Nothing was left. Unlike many women, I didn't need medical management to complete the process. The staff who dealt with me were polite, straightforward and quietly sympathetic.

Other women are not so lucky.

§ § §

Lizzie Lowrie has had six miscarriages, all in the first trimester. The care she has received has been patchy. She's met people who've been 'amazing', but she's also had to beg and cry down the phone to be admitted to hospital, and has turned up only to be congratulated on her pregnancy.

When she tells me about her most recent miscarriage, at ten weeks, I am shocked. She opted for a medically managed miscarriage, in which you take tablets that open the cervix to let the remaining tissue leave the body. 'It was horrendous,' she says. 'It was so painful. And I was in this ward with other people doing the same thing... It was terrible.'

Around 1 per cent of couples are affected by recurrent miscarriage, which in the UK is defined as the loss of three or more consecutive pregnancies.

Emma Benjamin has had several miscarriages, too, but still remembers the terror of the first. 'They just sent me home and they didn't tell me anything,' she says. 'I came home bleeding – having the most awful period pains, I suppose – and not really knowing what to do or what was going to happen or how long I was going to bleed for. I knew nothing, literally nothing... I wasn't given a leaflet or anything. So it was horrible, it was really awful, because I suppose I didn't really know what was going on.'

It's another side of the silence that surrounds miscarriage. But Benjamin and Lowrie both talk clearly and calmly about their experiences, and have become more open with each successive miscarriage. Lowrie tells me that for her husband and her the 12-week rule has 'gone out the window'.

'At first, very few people knew that I was pregnant,' she says, 'but then as the miscarriages went on we just made sure there were certain people close to us that knew... they tried to keep me sane when I was going through the pregnancy... It's just so hard breaking those two bits of news: I was pregnant and I'm not now. It's really hard to bring it into conversation.

'It is still quite a silent thing,' she adds, 'and I think part of it is that no one knows what to say.'

Coomarasamy agrees that lack of support is a serious problem and that women who have an early miscarriage, and their partners, may need just as much help as those who have lost an older baby to stillbirth. 'Whether it was this size baby or that size baby is irrelevant, and the psychological impact is not much different,' he explains. 'So I think there is a real need to understand how couples experience miscarriages. There's a real need to identify better ways of supporting the couples.'

Lowrie and her husband now run a blog about childlessness called Saltwater and Honey. Of course, no one should ever feel they have to share their experiences – I have friends who wanted to keep their miscarriage private. But it does seem that it's becoming increasingly acceptable to speak out about miscarriage.

§ § §

Breaking the silence is crucial. Research has shown that one-third of women attending specialist clinics as a result of their miscarriage are clinically depressed. As well as depression and grief, it's been reported that both women and their partners experience increased anxiety for several months after a miscarriage. Post-traumatic stress disorder, obsessive–compulsive disorders and panic disorders have also been observed in research studies.

Once, this would have surprised me. Not now. Three months after my own miscarriage, I still struggle to see my experience in perspective. There are still days when I feel a shadow over me and a sadness in the pit of my stomach that won't go away. There are still days when a strange emotion surprises me with its stranglehold.

It's only after my conversation with Lowrie that I realise this emotion is grief. She, too, was confused, until a counsellor demystified what she was going through.

'I thought to grieve you had to have lost something you'd met – like a person that you had talked to – or you could grieve over a baby that maybe you'd held,' she tells me. 'I didn't know anything about grief... I didn't know whether I should leave that to people who had lost actual people, not a very, very tiny baby that you've never met.'

Benjamin agrees: 'I used to think, 'God, people go through so much worse'... and I'd feel guilty for grieving... But in my head, I had planned when this baby was going to be born. So it was still as upsetting for me.'

Part of this distress comes from that unanswered 'Why?' Most women having their first or second miscarriage are told to put it down to one-off, unspecified genetic abnormalities in the fetus. It just wasn't meant to be. Yes, society likes fate. But women feel better if they get more accurate information, says Ruth Bender-Atik, national director of the Miscarriage Association. 'The reason is that they have an answer, an explanation,' she says, 'rather than a huge question mark and a tendency to assume it's their fault.'

Most women never get an answer, however, even if they are tested for possible explanations, because the science is sorely lacking.

'I think it's fair to say that miscarriage, despite being so common, despite having physical and psychological consequences to the woman and her partner, despite being a condition that demands quite a lot from the NHS, has not been researched well for a long time,' says Coomarasamy. 'But that is changing, I believe.'

§ § §

The unspecified genetic abnormalities that are said to underlie most miscarriages have various possible causes. The risk of random genetic faults in the fetus seems to increase with the age of the mother: the chances of having a miscarriage rise from 9 per cent aged 20–24 to more than 50 per cent for women aged 40 and over. Beyond age, other risk factors associated with miscarriage include obesity, smoking, drug use, and drinking more than two units of alcohol a week or more than a couple of cups of coffee a day.

There are several other potential causes: abnormalities in the womb or cervix, genetic faults inherited from the parents, hormone imbalances, polycystic ovary syndrome, various infections and so on. In the UK, tests for these possibilities are offered only after three consecutive miscarriages, whereas in many other countries the threshold is two.

Some women who've had a number of miscarriages have antibodies in their blood that seem to prevent the pregnancy embedding properly or cause blood clots in the placenta. This is called antiphospholipid syndrome, also commonly known as sticky blood syndrome, and it is the most important treatable cause of recurrent miscarriage. Low doses of aspirin, sometimes also the blood-thinning drug heparin, seem to help these women carry a pregnancy to term. It's the kind of hope many women and their partners cling to: that a cause will be found and an effective treatment will follow.

A blood test for these antibodies is therefore standard after recurrent miscarriages, but it's the answer only 15 per cent of the time. Half of all women who have tests are still left without an answer. Although Benjamin and her husband now have three children, a cause was never identified for her miscarriages. After two successful pregnancies in which she took progesterone, blood-thinning drugs, aspirin and steroids, she knows that it was more likely simply luck rather than targeted medical intervention.

Lowrie, still trying for her first child, has also tried taking low-dose aspirin, heparin and progesterone, but thinks she was probably only offered this cocktail of drugs because 'they just didn't know what to do with me'.

It's a familiar story to Coomarasamy. 'There are a lot of people out there who are just putting patients on a bit of this, a bit of that,' he says. 'Statistically speaking, any patient who has had a miscarriage previously – almost all patients who have had a miscarriage previously – the odds are in their favour in terms of having a normal pregnancy next time round, no matter what one does. So if they happen to be popping a pill it may have nothing to do with it. In fact, statistically speaking, they were going to carry that baby to term anyway.'

While aspirin increases the chances of a successful pregnancy for the minority of women with sticky blood syndrome, it had no significant effect in clinical trials for other women at risk of miscarrying. And following years of debate, the results of the PROMISE trial announced in November 2015 showed that progesterone supplements did not prevent early miscarriage for women with unexplained, recurrent losses.

A number of other trials continue to investigate potential treatments. The RESPONSE trial is testing a medicine called NT100 to find out if it can improve the chances of a successful pregnancy without serious side-effects. The TABLET trial is looking into the role that thyroid antibodies may play in women with unexplained miscarriage, and whether the drug levothyroxine may help. Lowrie is one of those waiting to hear if she is eligible to take part.

Lots of women seek out such trials, keen to be involved. Of Coomarasamy's patients at Birmingham, 60–70 per cent take part in clinical trials being carried out there, and often the research team finds recruits through other avenues, such as Facebook campaigns. They are all looking for answers, hoping for a breakthrough. But it may be that to understand miscarriage better, we need a new approach.

§ § §

Jan Brosens is a professor of obstetrics and gynaecology at the University of Warwick. He agrees that our current knowledge is too thin to help many people after recurrent miscarriage and says the current tests available are mostly a waste of time. '[For] the vast majority of couples that you see in clinic, you can test until you're blue in the face and you will find nothing,' he says. 'But more importantly, even if you have a patient where you have a positive test, you will find that same positive test in at least 50–100 women who don't have a history of miscarriages.'

In other words, the tests are nowhere near specific enough to identify what is causing recurrent miscarriages. Brosens thinks we will make more progress if we change the way we think about miscarriage.

'The problem I face when I see patients is that the vast majority come with this narrative that has been imposed upon them – and which they defend – which is that miscarriages are your body rejecting the pregnancy, that this is a complete failure,' he says, sadly.

Instead, he is keen to emphasise that a successful pregnancy begins with the start of a period – an event that so many women regard merely as an annoyance or, at worst, the uncomfortable end to another month of trying to conceive a baby.

But consider it differently, and the period is just the beginning, as the old womb lining disappears and a completely new one begins to grow. Brosens's research with Siobhan Quenby at Warwick's Biomedical Research Unit in Reproductive Health suggests that the womb lining plays a major role in determining whether the next pregnancy succeeds.

Most if not all human embryos have some chromosomal abnormalities. The range of variation runs from embryos with errors in a couple of cells right up to ones that are so unstable they are known as 'chaotic'. The cells of the womb lining, the endometrium, go through a process called decidualisation in response to the pregnancy hormone progesterone, which makes them able to recognise genetically poor embryos and prevent implantation so that pregnancy never begins.

But if the womb lining isn't suitably prepared, it may prevent healthy embryos from implanting – or do the opposite.

Brosens and Quenby's research has found that in women with recurrent miscarriage, the womb lining is often super-receptive but unselective, meaning that it allows genetically doomed embryos to implant and grow. These women may get pregnant fairly easily, but the pregnancy never truly has a chance of succeeding.

'In essence,' Brosens tells me as firmly as his friendly Dutch lilt allows, 'I completely and utterly dismiss the current view of miscarriages.'

The idea that something has 'gone wrong' in your pregnancy? No. The feeling of guilt that you must have done something wrong, despite sticking to all the rules of pregnancy? Pointless, because the outcome of your pregnancy was most likely determined at that moment of implantation.

Brosens is convinced that this new perspective will eventually lead to an uplifting advance: being able to predict who is at risk of miscarriages, even among women who have never been pregnant. When cells taken from the womb lining of women who have experienced recurrent miscarriage are cultured in the lab, 'the behaviour of the cells is very, very different [compared to] control patients,' he says. This provides a new starting-point for developing diagnostic tests and even treatments to make recurrent miscarriages far less likely.

§§§

Today, after just one miscarriage, the statistics tell me that I have an 80 per cent chance of my next pregnancy being successful. Regardless, I have been worrying that my miscarriage was the result of something that might make me prone to it happening again. I simply don't know, and it's the same for most women experiencing miscarriage, whether their first or their fifteenth.

The wonders of modern science have accustomed us to medical explanations and diagnoses. The women I've spoken to – Emma Benjamin, Lizzie Lowrie and some of my friends – share similar feelings of frustration. We expect that doctors will find out what is wrong with us and give us something to treat that problem. We think we will feel better if that happens.

For the small percentage of women whose every pregnancy has ended in miscarriage, the question of why looms particularly heavy over their trauma. While Lowrie hasn't given up hope of having a child of her own, she has accepted that it may not happen.

'I don't think there is always a resolution, but sometimes you've got to live with that,' she says. 'Life isn't neat. We don't always have answers.'

I don't have an answer, and I know I'm not going to get one any time soon. So for now, I'm going to try and stop asking 'Why?'

One tiny life has ended, but mine goes on.

This story was first published on 1 March 2016 by Wellcome on mosaicscience.com

Can the power of thought outwit ageing?

■ Jo Marchant

It's seven in the morning on the beach in Santa Monica, California. The low sun glints off the waves and the clouds are still golden from the dawn. The view stretches out over thousands of miles of Pacific Ocean. In the distance, white villas of wealthy Los Angeles residents dot the Hollywood hills. Here by the shore, curlews and sandpipers cluster on the damp sand. A few metres back from the water's edge, a handful of people sit cross-legged: members of a local Buddhist centre about to begin an hour-long silent meditation.

Such spiritual practices may seem a world away from biomedical research, with its focus on molecular processes and repeatable results. Yet just up the coast, at the University of California, San Francisco (UCSF), a team led by a Nobel Prize-winning biochemist is charging into territory where few mainstream scientists would dare to tread. Whereas Western biomedicine has traditionally shunned the study of personal experiences and emotions in relation to physical health, these scientists are placing state of mind at the centre of their work. They are engaged in serious studies hinting that meditation might – as Eastern traditions have long claimed – slow ageing and lengthen life.

§ § §

Elizabeth Blackburn has always been fascinated by how life works. Born in 1948, she grew up by the sea in a remote town in Tasmania, Australia, collecting ants from her garden and jellyfish from the beach. When she began her scientific career, she moved on to dissecting living systems molecule by molecule. She was drawn to biochemistry, she says, because it offered a thorough and precise understanding 'in the form of deep knowledge of the smallest possible subunit of a process'.

Working with biologist Joe Gall at Yale in the 1970s, Blackburn sequenced the chromosome tips of a single-celled freshwater creature called Tetrahymena ('pond scum', as she describes it) and discovered a repeating DNA motif that acts as a protective cap. The caps, dubbed telomeres, were subsequently found on human chromosomes too. They shield the ends of our chromosomes each time our cells divide and the DNA is copied, but they wear down with each division. In the 1980s, working with graduate student Carol Greider at the University of California, Berkeley, Blackburn discovered an enzyme called telomerase that can protect and rebuild telomeres. Even so, our telomeres dwindle over time. And when they get too short, our cells start to malfunction and lose their ability to divide – a phenomenon that is now recognised as a key process in ageing. This work ultimately won Blackburn the 2009 Nobel Prize in Physiology or Medicine.

In 2000, she received a visit that changed the course of her research. The caller was Elissa Epel, a postdoc from UCSF's psychiatry department. Psychiatrists and biochemists don't usually have much to talk about, but Epel was interested in the damage done to the body by chronic stress, and she had a radical proposal.

Epel, now director of the Aging, Metabolism and Emotion Center at UCSF, has a long-standing interest in how the mind and body relate. She cites as influences both the holistic health guru Deepak Chopra and the pioneering biologist Hans Selye, who first described in the 1930s how rats subjected to long-term stress become chronically ill. 'Every stress leaves an indelible scar, and the organism pays for its survival after a stressful situation by becoming a little older,' Selye said.

Back in 2000, Epel wanted to find that scar. 'I was interested in the idea that if we look deep within cells we might be able to measure the wear and tear of stress and daily life,' she says. After reading about Blackburn's work on ageing, she wondered if telomeres might fit the bill.

With some trepidation at approaching such a senior scientist, the then postdoc asked Blackburn for help with a study of mothers going through one of the most stressful situations that she could think of – caring for a chronically ill child. Epel's plan was to ask the women how stressed they felt, then look for a relationship between their state of mind and the state of their telomeres. Collaborators at the University of Utah would measure telomere length, while Blackburn's team would measure levels of telomerase.

Blackburn's research until this point had involved elegant, precisely controlled experiments in the lab. Epel's work, on the other hand, was on real, complicated people living real, complicated lives. 'It was another world as far as I was concerned,' says Blackburn. At first, she was doubtful that it would be possible to see any meaningful connection between stress and telomeres. Genes were seen as by far the most important factor determining telomere length, and the idea that it would be possible to measure environmental influences, let alone psychological ones, was highly controversial. But as a mother herself, Blackburn was drawn to the idea of studying the plight of these stressed women. 'I just thought, how interesting,' she says. 'You can't help but empathise.'

It took four years before they were finally ready to collect blood samples from 58 women. This was to be a small pilot study. To give the highest chance of a meaningful result, the women in the two groups – stressed mothers and controls – had to match as closely as possible, with similar ages, lifestyles and backgrounds. Epel recruited her subjects with meticulous care. Still, Blackburn says, she saw the trial as nothing more than a feasibility exercise. Right up until Epel called her and said, 'You won't believe it.'

The results were crystal clear. The more stressed the mothers said they were, the shorter their telomeres and the lower their levels of telomerase.

The most frazzled women in the study had telomeres that translated into an extra decade or so of ageing compared to those who were least stressed, while their telomerase levels were halved. 'I was thrilled,' says Blackburn. She and Epel had connected real lives and experiences to the molecular mechanics inside cells. It was the first indication that feeling stressed doesn't just damage our health – it literally ages us.

§ § §

Unexpected discoveries naturally meet scepticism. Blackburn and Epel struggled initially to publish their boundary-crossing paper. 'Science [one of the world's leading scientific journals] couldn't bounce it back fast enough!' chuckles Blackburn.

When the paper finally was published, in the Proceedings of the National Academy of Sciences in December 2004, it sparked widespread press coverage as well as praise. Robert Sapolsky, a pioneering stress researcher at Stanford University and author of the bestselling Why Zebras Don't Get Ulcers, described the collaboration as 'a leap across a vast interdisciplinary canyon'. Mike Irwin, director of the Cousins Center for Psychoneuroimmunology at the University of California, Los Angeles, says it took a lot of courage for Epel to seek out Blackburn. 'And a lot of courage for Liz [Blackburn] to say yes.'

Many telomere researchers were wary at first. They pointed out that the study was small, and questioned the accuracy of the telomere length test used. 'This was a risky idea back then, and in some people's eyes unlikely,' explains Epel. 'Everyone is born with very different telomere lengths and to think that we can measure something psychological or behavioural, not genetic, and have that predict the length of our telomeres? This is really not where this field was ten years ago.'

The paper triggered an explosion of research. Researchers have since linked perceived stress to shorter telomeres in healthy women as well as in Alzheimer's caregivers, victims of domestic abuse and early life trauma, and people with major depression and post-traumatic stress disorder. 'Ten years on, there's no question in my mind that the environment has some consequence on telomere length,' says Mary Armanios, a clinician and geneticist at Johns Hopkins School of Medicine who studies telomere disorders.

There is also progress towards a mechanism. Lab studies show that the stress hormone cortisol reduces the activity of telomerase, while oxidative stress and inflammation – the physiological fallout of psychological stress – appear to erode telomeres directly.

This seems to have devastating consequences for our health. Age-related conditions from osteoarthritis, diabetes and obesity to heart disease, Alzheimer's and stroke have all been linked to short telomeres.

The big question for researchers now is whether telomeres are simply a harmless marker of age-related damage (like grey hair, say) or themselves play a role in causing the health problems that plague us as we age. People with genetic mutations affecting the enzyme telomerase, who have much shorter telomeres than normal, suffer from accelerated-ageing syndromes and their organs progressively fail. But Armanios questions whether the smaller reductions in telomere length caused by stress are relevant for health, especially as telomere lengths are so variable in the first place.

Blackburn, however, says she is increasingly convinced that the effects of stress do matter. Although the genetic mutations affecting the maintenance of telomeres have a smaller effect than the extreme syndromes Armanios studies, Blackburn points out that they do increase the risk of chronic disease later in life. And several studies have shown that our telomeres predict future health. One showed that elderly men whose telomeres shortened over two-and-a-half years were three times as likely to die from cardiovascular disease in the subsequent nine years as those whose telomeres stayed the same length or got longer. In another study, looking at over 2,000 healthy Native Americans, those with the shortest telomeres were more than twice as likely to develop diabetes over the next five-and-a-half years, even taking into account conventional risk factors such as body mass index and fasting glucose.

Blackburn is now moving into even bigger studies, including a collaboration with healthcare giant Kaiser Permanente of Northern California that has involved measuring the telomeres of 100,000 people. The hope is that combining telomere length with data from the volunteers' genomes and electronic medical records will reveal additional links between telomere length and disease, as well as more genetic mutations that affect telomere length. The results aren't published yet, but Blackburn is excited about what the data already shows about longevity. She traces the curve with her finger: as the population ages, average telomere length goes down. This much we know; telomeres tend to shorten over time. But at age 75–80, the curve swings back up as people with shorter telomeres die off – proof that those with longer telomeres really do live longer. 'It's lovely,' she says. 'No one has ever seen that.'

In the decade since Blackburn and Epel's original study, the idea that stress ages us by eroding our telomeres has also permeated popular culture. In addition to Blackburn's many scientific accolades, she was named one of Time magazine's '100 most influential people in the world' in 2007, and received a Good Housekeeping achievement award in 2011. A workaholic character played by Cameron Diaz even described the concept in the 2006 Hollywood film The Holiday. 'It resonates,' says Blackburn.

But as evidence of the damage caused by dwindling telomeres piles up, she is embarking on a new question: how to protect them.

§§§

At first, the beach seems busy. Waves splash and splash and splash. Sanderlings wheel along the shoreline. Joggers and dog walkers amble across, while groups of pelicans hang out on the water before taking wing or floating out of sight. A surfer, silhouetted black against the sky, bobs about for 20 minutes or so, catching the odd ripple towards shore before he, too, is gone. The unchanging perspective gives a curious sense of detachment. You can imagine that the birds and joggers and surfers are like thoughts: they inhabit different forms and timescales but in the end, they all pass.

There are hundreds of ways to meditate but this morning I'm trying a form of Buddhist mindfulness meditation called open monitoring, which involves paying attention to your experience in the present moment. Sit upright and still, and simply notice any thoughts that arise – without judging or reacting to them – before letting them go. For Buddhists this is a spiritual quest; by letting trivial thoughts and external influences fall away, they hope to get closer to the true nature of reality.

Blackburn too is interested in the nature of reality, but after a career spent focusing on the measurable and quantifiable, such navel-gazing initially held little personal appeal and certainly no professional interest. 'Ten years ago, if you'd told me that I would be seriously thinking about meditation, I would have said one of us is loco,' she told the New York Times in 2007. Yet that is where her work on telomeres has brought her. Since her initial study with Epel, the pair have become involved in collaborations with teams around the world – as many as 50 or 60, Blackburn estimates, spinning in 'wonderful directions'. Many of these focus on ways to protect telomeres from the effects of stress; trials suggest that exercise, eating healthily and social support all help. But one of the most effective interventions, apparently capable of slowing the erosion of telomeres – and perhaps even lengthening them again – is meditation.

So far the studies are small, but they all tentatively point in the same direction. In one ambitious project, Blackburn and her colleagues sent participants to meditate at the Shambhala mountain retreat in northern Colorado. Those who completed a three-month course had 30 per cent higher levels of telomerase than a similar group on a waiting list. A pilot study of dementia caregivers, carried out with UCLA's Irwin and published in 2013, found that volunteers who did an ancient chanting meditation called Kirtan Kriya, 12 minutes a day for eight weeks, had significantly higher telomerase activity than a control group who listened to relaxing music. And a collaboration with UCSF physician and self-help guru Dean Ornish, also published in 2013, found that men with low-risk prostate cancer who undertook comprehensive lifestyle changes, including meditation, kept their telomerase activity higher than similar men in a control group and had slightly longer telomeres after five years.

In their latest study, Epel and Blackburn are following 180 mothers, half of whom have a child with autism. The trial involves measuring the women's stress levels and telomere length over two years, then testing the effects of a short course of mindfulness training, delivered with the help of a mobile app.

Theories differ as to how meditation might boost telomeres and telomerase, but most likely it reduces stress. The practice involves slow, regular breathing, which may relax us physically by calming the fight-or-flight response. It probably has a psychological stress-busting effect too. Being able to step back from negative or stressful thoughts may allow us to realise that these are not necessarily accurate reflections of reality but passing, ephemeral events. It also helps us to appreciate the present instead of continually worrying about the past or planning for the future.

'Being present in your activities and in your interactions is precious, and it's rare these days with all of the multitasking we do,' says Epel. 'I do think that in general we've got a society with scattered attention, particularly when people are highly stressed and don't have the resources to just be present wherever they are.'

§ § §

Inevitably, when a Nobel Prize-winner starts talking about meditation, it ruffles a few feathers. In general, Blackburn's methodical approach to the topic has earned a grudging admiration, even among those who have expressed concern about the health claims made for alternative medicine. 'She goes about her business in a cautious and systematic fashion,' says

Edzard Ernst of the University of Exeter, UK, who specialises in testing complementary therapies in rigorous controlled trials. Oncologist James Coyne of the University of Pennsylvania, Philadelphia, who is sceptical of this field in general and describes some of the research on positive psychology and health as 'morally offensive' and 'tooth fairy science', concedes that some of Blackburn's data is 'promising'.

Others aren't so impressed. Surgeon-oncologist David Gorski is a well-known critic of alternative medicine and pseudoscience who blogs under the name of Orac – he's previously described Dean Ornish as 'one of the four horsemen of the Woo-pocalypse'. Gorski stops short of pronouncing meditation as off-limits for scientific inquiry, but expresses concern that the preliminary results of these studies are being oversold. How can the researchers be sure they're investigating it rigorously? 'It's really hard to do with these things,' he says. 'It is easy to be led astray. Nobel Prize-winners are not infallible.' Blackburn's own biochemistry community also seems ambivalent about her interest in meditation. Three senior telomere researchers I contacted declined to discuss this aspect of her work, with one explaining that he didn't want to comment 'on such a controversial issue'.

'People are very uncomfortable with the concept of meditation,' notes Blackburn. She attributes this to its unfamiliarity and its association with spiritual and religious practices. 'We're always trying to say it as carefully as we can... always saying 'look, it's preliminary, it's a pilot'. But people won't even read those words. They'll see the newspaper headings and panic.'

Any connotation of religious or paranormal beliefs makes many scientists uneasy, says Chris French, a psychologist at

Goldsmiths, University of London, who studies anomalous experiences including altered states of consciousness. 'There are a lot of raised eyebrows, even though I've got the word sceptic virtually tattooed across my forehead,' he says. 'It smacks of new-age woolly ideas for some people. There's a kneejerk dismissive response of 'we all know it's nonsense, why are you wasting your time?''

'When meditation first came to the West in the 1960s it was tied to the drug culture, the hippie culture,' adds Sara Lazar, a neuroscientist at Harvard who studies how meditation changes the structure of the brain. 'People think it's just a bunch of crystals or something, they roll their eyes.' She describes her own decision to study meditation, made 15 years ago, as 'brave or crazy', and says that she only plucked up the courage because at around the same time, the US National Institutes of Health (NIH) created the National Center for Complementary and Alternative Medicine. 'That gave me the confidence that I could do this and I would get funding.'

The tide is now turning. Helped in part by that NIH money, researchers have developed secularised – or non-religious – practices such as mindfulness-based stress reduction and mindfulness-based cognitive therapy, and reported a range of health effects from lowering blood pressure and boosting immune responses to warding off depression. And the past few years have seen a spurt of neuroscience studies, like Lazar's, showing that even short courses of meditation can forge structural changes in the brain.

'Now that the brain data and all this clinical data are coming out, that is starting to change. People are a lot more accepting [of meditation],' says Lazar. 'But there are still some people who will never believe that it has any benefit whatsoever.'

Blackburn's view is that meditation is a fair topic to study, as long as robust methods are used. So when her research first pointed in this direction, she was undaunted by concerns about what such studies might do to her reputation. Instead, she tried it out for herself, on an intensive six-day retreat in Santa Barbara. 'I loved it,' she says. She still uses short bursts of meditation, which she says sharpen her mind and help her to avoid a busy, distracted mode. She even began one recent paper with a quote from the Buddha: 'The secret of health for both mind and body is not to mourn for the past, worry about the future, or anticipate troubles but to live in the present moment wisely and earnestly.'

That study, of 239 healthy women, found that those whose minds wandered less – the main aim of mindfulness meditation – had significantly longer telomeres than those whose thoughts ran amok. 'Although we report merely an association here, it is possible that greater presence of mind promotes a healthy biochemical milieu and, in turn, cell longevity,' the researchers concluded. Contemplative traditions from Buddhism to Taoism believe that presence of mind promotes health and longevity; Blackburn and her colleagues now suggest that the ancient wisdom might be right.

§§§

I meet with Blackburn in Paris. We're at an Art Nouveau-themed bistro just down the road from the Curie Institute, where she is on a short sabbatical, arranging seminars between groups of scientists who don't usually talk to one another. In a low, melodious voice that I strain to hear through the background clatter, the 65-year-old tells me of her first major brush with Buddhist thinking.

In September 2006, she attended a conference held at the Menla Mountain Buddhist centre, a remote retreat in New York's Catskill mountains, at which Western scientists met with Tibetan-trained scholars including the Dalai Lama to discuss longevity, regeneration and health. During the meeting, the spiritual leader honoured Blackburn's scientific achievements by inducting her as a 'Medicine Buddha'.

If Epel's psychiatry research had been another world, the scholars' Eastern philosophy seemed to Blackburn more alien still. Over dinner one evening, while explaining to the other delegates how errors in the gene for telomerase can cause health problems, she described genetic mutation as a random, chance event. That's dogma for Western scientists but not for those trained in the Tibetan worldview. 'They said 'oh no, we don't regard this as chance',' says Blackburn. For these holistic scholars, even the smallest events were infused with meaning. 'I suddenly thought, whoa, this is a very different world from the one I'm on.'

But instead of dismissing her Eastern counterparts, she was impressed, finding the Dalai Lama to have 'a very good brain', for example. 'They're scholarly in a very different way, but it is still good-quality thinking,' she explains. 'It wasn't 'God told me this', it was more 'let's see what actually happens in the brain'. So there are certain elements of the approach that I am quite comfortable with as a scientist.'

Blackburn isn't tempted to embrace the spiritual approach herself. 'I'm rooted in the physical world,' she says. But she combines that grounding with an open mind towards new ideas and connections, and she seems to love breaking out of established paradigms. For example, she and Epel have shown that the effects of stress on telomeres can be passed on to the next generation. If women experience stress while pregnant, their children have shorter telomeres, as newborns and as adults – in direct contradiction of the standard view that traits can only be passed on via our genes.

In the future, information from telomeres may help doctors decide when to prescribe particular drugs. For example, telomerase activity predicts who will respond to treatment for major depression, while telomere length influences the effects of statins. In general, however, Blackburn is more interested in how telomeres might help people directly, by encouraging them to live in a way that reduces their disease risk. 'This is not a familiar model for the medical world,' she says.

Conventional medical tests give us our risk of particular conditions – high cholesterol warns of impending heart disease, for example, while high blood sugar predicts diabetes. Telomere length, by contrast, gives an overall reading of how healthy we are: our biological age. And although we already know that we should exercise, eat well and reduce stress, many of us fall short of these goals. Blackburn believes that putting a concrete number on how we are doing could provide a powerful incentive to change our behaviour. In fact, she and Epel have just completed a study (as yet unpublished) showing that simply being told their telomere length caused volunteers to live more healthily over the next year than a similar group who weren't told.

Ultimately, however, the pair want entire countries and governments to start paying attention to telomeres. A growing body of work now shows that the stress from social adversity and inequality is a major force eroding these protective caps. People who didn't finish high school or are in an abusive relationship have shorter telomeres, for example, while studies have also shown links with low socioeconomic status, shift work, lousy neighbourhoods and environmental pollution. Children are particularly at risk: being abused or experiencing adversity early in life leaves people with shorter telomeres for the rest of their lives. And through telomeres, the stress that women experience during pregnancy affects the health of the next generation too, causing hardship and economic costs for decades to come.

In 2012, Blackburn and Epel wrote a commentary in the journal Nature, listing some of these results and calling on politicians to prioritise 'societal stress reduction'. In particular, they argued, improving the education and health of women of child-bearing age could be 'a highly effective way to prevent poor health filtering down through generations'. Meditation retreats or yoga classes might help those who can afford the time and expense, they pointed out. 'But we are talking about broad socioeconomic policies to buffer the chronic stressors faced by so many.' Where many scientists refrain from discussing the political implications of their work, Blackburn says she wanted to speak out on behalf of women who lack support, and say 'You'd better take their situations seriously.'

While arguments for tackling social inequality are hardly new, Blackburn says that telomeres allow us to quantify for the first time the health impact of stress and inequality and therefore the resulting economic costs. We can also now pinpoint pregnancy and early childhood as 'imprinting periods' when telomere length is particularly susceptible to stress. Together, she says, this evidence makes a stronger case than ever before for governments to act.

But it seems that most scientists and politicians still aren't ready to leap across the interdisciplinary canyon that Blackburn and Epel bridged a decade ago. The Nature article has engendered little response, according to a frustrated Epel. 'It's a strong statement so I would have thought that people would have criticised it or supported it,' she says. 'Either way!'

'It's now a consistent story that the ageing machinery is shaped at the earliest stages of life,' she insists. 'If we ignore that and we just keep trying to put band-aids on later, we're never going to get at prevention and we're only going to fail at cure.' Simply responding to the physical symptoms of disease might make sense for treating an acute infection or fixing a broken leg, but to beat chronic age-related conditions such as diabetes, heart disease and dementia, we will need to embrace the fuzzy, subjective domain of the mind.

This story was first published on 1 July 2014
by Wellcome on mosaicscience.com

Seeking a 'cure' for male baldness

■ Rhodri Marsden

When I was a teenager, my mum reassured me that I wouldn't go bald in my 20s like my father had. She seemed pretty sure of this. 'Look,' she said, brightly, pushing back her hair from her forehead, 'you've got my hairline, not your dad's.' At the time, I bought her argument, but within 10 years her reasoning had been revealed to be magnificently wrong. My hairline had begun its slow march north, a clear sign that I'd inherited male pattern hair loss from one of my parents, if not both of them.

I pretended not to be bothered at the time, and as the years went by I persuaded myself that my ever-shorter haircuts might make me look better than I used to. But deep down it felt unfair, a genetic quirk I didn't deserve. I'd wince as the barber held up the mirror behind me, revealing an ever-widening bald patch. The advent of social media gave me an exciting new pastime: untagging myself from photos that gave an unflattering perspective on my gleaming forehead, which was pretty much all of them. I pretended not to be bothered, but it was a pretence, and that pretence continues today, in my 40s; while stoically accepting hair loss as my destiny, I know perfectly well how I feel about it. I don't like it. I've found myself turning to classic coping mechanisms such as wearing hats and growing a beard, hilariously feeble attempts at misdirection that fool nobody, least of all me.

'Hair are your aerials,' says Danny, the hirsute drug dealer in the film Withnail and I. 'They pick up signals from the cosmos and transmit them directly into the brain. This is the reason bald-headed men are uptight.' Danny's conclusion – that all hairdressers are in the employment of the government – was stoner paranoia, but in one sense he was right: balding men are often uptight, about their baldness if nothing else. As coping mechanisms go, my hat and my beard are pretty benign examples. But while hats and beards tend not to provoke any additional anxiety, it would seem that transplants, drugs and wigs certainly do.

Androgenetic alopecia is the medical term for this inherited form of hair loss, and while it affects both men and women, it's men whose anxieties tend to be targeted by the hair loss industry. It's estimated to be worth at least $1.5 billion a year worldwide, servicing the needs of millions of men and becoming increasingly adept at persuading them to part with money. A casual internet search returns a disorientating array of options that promise to alleviate the misery of the balding man: from herbal remedies to surgical procedures, from magic foams to fancy hairpieces, from restorative shampoos to nanofibre sprays for 'colouring in' bald patches. Some of them work, in the sense that the hair loss might be made less apparent (no miracle cures exist), but what works for one person might pan out disastrously for another. The resulting arguments play out daily between thousands of voices across dozens of websites, helping to generate a smokescreen of confusion behind which snake-oil salesmen can operate freely.

Spencer Stevenson started losing his hair at a young age, and he's spoken widely in the media and online about the trauma it has caused him. To assuage his misery he ended up having treatments costing a total of £40,000, including 11 hair transplants, many of which fell way short of his initial expectations. Following that horrific experience, he's become a vocal mentor for those suffering from hair loss, offering advice and detailing his suffering at the hands of what he considers to be a brutal, cut-throat industry. 'This is the problem,' he says. 'It's governed by money, and there are only a few organisations that have the patient's best interests at heart. The industry has an ugly reputation for preying on the vulnerable.'

This vulnerability is rarely acknowledged, but it's widespread. A 2005 study, spanning five European countries, found that 43 per cent of men with hair loss were concerned about its effect on personal attractiveness, with 22 per cent worrying about its impact on their social life and 21 per cent linking it with feelings of depression. Alopecia areata, an autoimmune disease that causes hair loss in men and women, has its own psychological consequences, which are often discussed in the media, but the incredibly widespread condition of male pattern hair loss has caused distress since the year dot and is talked about far less. History tells us of men willing to try all manner of bizarre remedies to thwart it, while bystanders find their frustration (and indeed their hair loss) faintly amusing. In the Old Testament, the prophet Elisha is taunted for his baldness by a group of boys as he heads to Bethel; he's sufficiently touchy about this to call for the assistance of God, who promptly summons two bears to maul the boys to death. Harsh, certainly – but it's worth noting that God chose to exterminate the taunters rather than tackle the hair loss. You can't really blame Him, though. Male pattern hair loss is a very tricky problem indeed.

§ § §

According to the UK's National Institute for Health and Care Excellence, the condition affects 30 per cent of men under 30, increasing to around 80 per cent of men over 70. Its causes are well established, but poorly understood by those of us who have it. We might blame blocked pores, over-shampooing, over-brushing, the water supply or even the remedies we've bought, but the truth is that it's a cruel trick played by nature on the genetically susceptible. Dihydrotestosterone (DHT) is thought to be the hormone responsible; it's synthesised from testosterone by an enzyme, 5-alpha-reductase, that's found in the dermal papilla, a small compartment at the base of the hair follicle. This kicks off a process of miniaturisation in hormonally sensitive areas such as the forehead and the crown. The dermal papilla cells reduce in number, the follicles shrink and, as the American Hair Loss Association puts it, they stop producing 'cosmetically acceptable hair'.

The first consequence of this is progressive baldness. The second, and arguably more important, is our psychological response to it. 'It's whatever it means for that individual man,' says Anthony Bewley of the British Association of Dermatologists, who has a special interest in the psychology of skin conditions. 'A sense of loss of attractiveness, a loss of youth, a loss of virility, or even emasculation. And although it's a physiological change, ie something that happens as you get older, to dismiss it as something that isn't a disease, or something that doesn't matter, or something that's just your hair – that's utterly unhelpful for people whose confidence is compromised.'

A balding Bruce Willis is widely considered to be sexy, but this isn't enough to reassure most of us that our thinning hair isn't a curse on our masculinity. For every relationship that breaks down, falters or fails to begin, baldness is frequently used as a scapegoat. 'If only I had a full head of hair,' goes the train of thought, 'things would be different.' The logic is hilariously flawed – but it's a seed that can grow quickly if it isn't kept in check, and is fertilised by a culture that encourages bald men to conceal their condition, often leading to repressed anger, unhappiness and resentment. Most healthcare professionals would agree that coming to terms with hair loss would be the best option for most men, but this avenue is rarely explored by those in distress; most of us see the problem to be tackled as hair loss, rather than our attitude towards it.

'I'd say that the most desperate emails, the people who sound like they're at their absolute lowest ebb, tend to be from men,' says Amy Johnson at Alopecia UK, a charity offering support and advice to men and women with all types of alopecia. 'By the point they're getting in touch with us, they've not felt they've been able to speak to anyone else about their feelings, so it's an outpouring to an anonymous person. But when people say it's much harder for women, and for men it's alright, I say actually, no, that's not what I find from the support emails that come in.'

§ § §

Jay Patel, the co-founder of MH2Go, a wig supply and fitting business, sits in his office just off Brick Lane in central London, fiddling with a pen as he recounts his tale of hair loss. 'About four or five years ago I tried to commit suicide,' he says. 'There was a lot of other stuff going on, because I also suffer from body dysmorphia. I was put in hospital for three weeks, and I got a lot of support. After that I told everyone that I wore a wig, and the whole burden just lifted. I stopped feeling embarrassed, I stopped feeling ashamed.' I can't help but glance up at Patel's hairline; you'd never tell that it's a wig. He's a good-looking chap, and you sense that he'd look great with or without hair. But having told me about his wig, he forces a smile. He knows that I know. 'I've turned something that was an issue into my livelihood,' he says. MH2Go grew from a bedroom-based mail-order service, MensHair2Go, into a more discreetly named high-street operation. From the street the premises look like a regular salon, but inside Patel offers advice to people looking to buy wigs, while his business partner, Egita Rogule, styles and fits them. The price list is, by wig standards, very reasonable: £495 for the first and £250 for each subsequent 'system', each of which should last between four and six months.

Now it's Patel's turn to look at my forehead. 'You see,' he says, 'you wouldn't be a good hair transplant client because of the larger area to cover. It's just not feasible.' He picks out a hairpiece from a box. 'So, we call this one the iBase,' he says. 'All companies give them their own names – you know, Super, Elite and so on – but they're all the same, unprocessed human hair. Do you want to try one? You probably should, as you're here.' I'd already decided before my visit that I didn't want to have a fitting, that I didn't want to emulate Elton John or, for that matter, Burt Reynolds. But hey, it's just me and Jay, so I bite the bullet and sit down in front of a mirror. 'So, this isn't made for your head shape, or anything,' he says, 'and it's black hair, so it's not your colour. Just imagine that it's grey.' Patel puts it into position and stands back. 'Actually,' he says, 'it looks quite good.' I'm inclined to agree with him – but at the end of the day, it's a wig. I'm wearing a bloody wig.

'I do my consultations in a way that makes people aware of what they're getting themselves into,' says Patel. 'I say, look, it's a wig. It's never going to be your own hair. We'll make it look the best we can, but you're still wearing a wig.' His honest approach stems from a bitter experience he had in his early 20s, when he paid a company £20,000 for a series of wigs that lasted just a few weeks. 'I ran out of money, because I was only 23,' he says. 'But I was locked in with them. It's like a drug, they were my suppliers.' The nature of wig fitting, which requires the top of the head to be shaved before it's glued or taped, means that once you're in the game it's not easy to get out. 'So you can take advantage of clients,' explains Patel. 'If you go to other companies, you'll be met with people who are very cocksure, very confident, because they're in it to sell, sell, sell. It's very rare to meet someone in this industry who's been what I've been through.'

§§§

Nadeem Uddin Khan, director of the Harley Street Hair Clinic, is about to disprove Patel's theory. 'I'm just looking for a picture of myself, somewhere here,' he says, flicking through his phone before finally finding what he's looking for. He turns it around to show me: it's a picture of a balding guy, good-looking, perhaps slightly self-conscious. Again, I instinctively look back up at his current hairline: it's neat, buzzcut and very different from the photo. 'That was about 10, 12 years ago,' he says, putting his phone face down on the table. 'When I lost my hair it was just terrible. A year of not going out. So I understand what these guys go through.' Khan was one of the first people in the UK to undergo the FUE (follicular unit extraction) method of hair transplantation, which his clinic now specialises in. FUE is regarded, particularly by younger men, as a successful, modern surgical procedure with minimal stigma. This may be largely down to footballer Wayne Rooney's two hair transplants, both performed at Khan's clinic. 'He's been a great ambassador for us, and for hair transplantation in general,' says Khan. 'It's led to a massive spike of interest from all around the world.'

After donning some protective clothing, I'm led into a surgery, where a man is lying on his back, arms folded, as a doctor uses a special tool to make incisions in his forehead. He spent the morning lying on his front as follicles were extracted from the back of his head; later they'll be popped into their new location. He confesses that he's a bit bored, but he's looking forward to the results. It's his second op; he was so delighted with the results of the first that he's come back for another. Rooney, again, turns out to have been the catalyst. 'The thing is,' he says cheerily, as the surgeon swabs blood from his scalp, 'with Rooney you're not seeing photographs of his head taken from special angles and with special lighting. You're seeing him running around a football field, sweating, on the telly, every week. The results are there for everyone to see.'

§§§

All hair transplants are based on the principle of donor dominance, developed in the 1950s by New York dermatologist Norman Orentreich: a transplanted follicle doesn't know it's been moved; it just continues to grow as if it had been left where it was. Follicles taken from the back and the sides of the head – areas that aren't sensitive to the miniaturisation caused by the hormone DHT – 'remember' their lack of sensitivity when transplanted into bald areas. By the 1980s, experiments with grafting had developed into a procedure known as FUT (follicular unit transplantation) or strip surgery. A strip of the scalp is removed and cut into very small segments, which are then inserted into small holes in the affected area. It's a quicker procedure than FUE, and is considered by some to be the best way of harvesting high-quality hair. However, it leaves a long scar where the strip has been removed. This scarring, along with a number of poor-quality procedures undergone by high-profile celebrities, has given FUT a bad reputation that may not be entirely deserved.

FUE clinics tend to be quick to denigrate FUT, but FUE also has its drawbacks. As I witnessed at the Harley Street

Hair Clinic, it's a gruelling process that requires great concentration and stamina on the part of the physician, and enormous patience from the patient. Thousands of individual follicles are selected from across the donor area, extracted with a special tool, kept chilled, and later transplanted into tiny incisions. 'The procedure for me – around 3,000 grafts – was so labour-intensive that it took all day, from 8.30am to 5.30pm,' says one man who recently underwent an FUE transplant in Australia. 'The surgeon gave the sense of approaching her work with an artistic feel, in terms of ensuring consistency with the head's whorl and general density.'

This is the key to a good FUE transplant, according to Khan. 'The transplants should be done by a physician who knows what they're doing,' he says. 'The right follicle has to go in the right place. Mother Nature doesn't work in straight lines. These little design details are critical.' Khan shows me the first post-op photographs of the man whose second procedure I just witnessed. They're great. If I had a spare £10,000, which I don't, and if I had a bigger area of donor hair, which I don't either, I wonder if I might be persuaded to take the plunge. On balance, the answer is probably yes.

Sadly, not everyone performing these operations pays such a close attention to detail. 'There are tens of thousands of clinics,' says Spencer Stevenson. 'Often you have nurses performing surgery in the back of a dentist's! It's unbelievable what is going on. In Turkey there's a conveyor belt of patients because it's so cheap, but I've always said that money and geography should be the last considerations. As difficult as it is to go and have surgery, trust me, it's a thousand times more difficult once you become a repair patient.'

Given the risks associated with dodgy clinics, why on earth do men put themselves through this? The sense that a head of hair bestows us with greater masculinity and sexual appeal is evidently very deep-seated, and it's not something the industry has ever rushed to dispel. In June of this year the Farjo Hair Institute, an acclaimed British FUE transplant clinic which recently developed a pioneering robot (ARTAS) to perform hair extractions, released the results of an experiment it had done with a patient using the dating app Tinder. The results, breathlessly reported by the media, showed that the patient's post-transplant pictures resulted in 75 per cent more matches than pre-transplant. There was little scientific rigour in this, but men only seem to require a small amount of anecdotal evidence to reinforce their belief that male pattern hair loss is inherently unattractive.

'I used to have really thick hair,' says Jay Patel, wistfully. 'Especially in our culture, South Asian culture, if you don't have hair, you're not going to get marriage proposals. You see all your friends and your cousins, they all have amazing hair, but you're suffering from hair loss.' Patel tells heartbreaking stories of clients who arrive at his shop who are clearly traumatised. 'One guy spent half an hour standing outside because he was too scared to come in and even discuss hair loss,' he said. 'He eventually came in, wearing a bandana. I asked him to take it off to assess him, but he was really reluctant. He wanted to make sure the door was closed. Then he asked who was going to style his hair. I said it'll be my colleague Egita, and he said that he didn't want a girl to see him like this.'

§ § §

The clinical psychologist Paul Gilbert argued in a 1997 paper that our drive to be perceived as attractive is innate, and feelings of shame and humiliation result from negative reactions as we ponder whether those reactions are deserved or not. Much of the money that men spend on attempts to cure, mask or lessen their hair loss is evidently to do with that shame – despite male pattern hair loss having far less stigma than, say, a more serious or unusual skin condition. And yet many men seem unable to take that on board, particularly in the modern age, when those negative reactions are easily experienced via dating apps and regularly reinforced by the media, who often refer to the 'misery', 'suffering' and 'hope' of those with the condition. 'The reality of Western culture,' says consultant dermatologist Anthony Bewley, 'is that young is beautiful, more successful, more attractive. Sadly, if you are older, uglier, scarred and bald, it's not just about what that means in terms of your ability to succeed or be attractive; it's also, sadly, marked out as being less worthy – or even criminal.'

Faced with three options that all have disadvantages – the stigma of wig-wearing, invasive surgical procedures, or doing nothing whatsoever – a comparatively easy way of treating male pattern hair loss might seem to be drugs. From the reams of sales pitches and glowing testimonials for various foams, sprays, creams and tablets, you'd think that among them there would be at least one drug, one magic solution that would make hair miraculously re-sprout from a bald pate. But this isn't the case. There are just two approved drugs on the market, minoxidil and finasteride, and it's generally agreed that neither can reverse hair loss but merely delay the inevitable.

The link between minoxidil and hair growth was first noticed in the 1960s by men who took it during clinical trials for a treatment for high blood pressure. The link is still not fully understood, but the US Food and Drug Administration approved it as a hair loss treatment for men in 1988 with the proviso that it 'will not work for everyone'. Trials have shown some form of regrowth in as many as 80 per cent of patients, but a 2015 systematic review of the efficacy of minoxidil, while acknowledging that the medication was 'more effective than placebo in promoting total and nonvellus hair growth', also stated that 'cosmetically acceptable results are present in only a subset of patients'. Minoxidil is available as an over-the-counter topical medication under such promising-sounding names as Hair Grow, Hairgain, Hairway and Splendora; in the UK and USA it's usually known as Rogaine or Regaine. 'It's a bit of a Band-Aid,' says Stevenson. 'It's a good add-on as part of your regime, but on its own it's not going to get you too far.'

There are just two approved drugs on the market, and it's generally agreed that neither can reverse hair loss

Finasteride, meanwhile, has been available in the USA for hair growth under the name Propecia since 1997. Its hair-growing properties were first noticed by users of Proscar, a 5 mg dose of finasteride originally manufactured by Merck for treating benign prostatic hyperplasia, or enlarged prostate. After tests, Merck determined that a 1 mg dose was sufficient to promote hair growth; Propecia is thus effectively just one-fifth of a Proscar tablet (indeed, many men save money on prescriptions by cutting Proscar tablets into pieces). Its workings are better understood than minoxidil's (it's thought to inhibit the 5-alpha-reductase enzyme in the dermal papilla) but some of the side-effects listed by the US Food and Drug Administration (FDA), including erectile dysfunction, libido disorders, ejaculation disorders and orgasm disorders, can put men off. 'There are millions of people taking finasteride with no side-effects,' says a disgruntled Stevenson, 'but you don't hear from them. All you hear is the minority of people who have, and they put the fear of God in everybody. I feel sorry for the people who are scared.'

No other medications for male pattern hair loss are licensed by the FDA or its European equivalent, the European Medicines Agency, but there are many dietary supplements with medical-sounding names that claim to promote healthy hair growth. While there may be a link between poor nutrition and some forms of hair loss, there's no mention of dietary supplements in the NICE guidance on androgenetic alopecia. Nevertheless, the relative merits of supplements such as TRX2 and Viviscal are vigorously debated online.

While those arguments continue, the pursuit of a wonder drug continues, with dozens of companies keen to reap the financial rewards. One such company, Allergan, has two potentially promising drugs undergoing trials: a topical one, bimatoprost, which was originally a treatment for glaucoma and was approved by the FDA in 2008 to assist with the growth of eyelashes; and an oral medication, setipiprant, which inhibits a compound (prostaglandin D2) that's found to be elevated in balding scalps. Another company, Samumed, has recently picked up a huge amount of press interest for another potential treatment, SM04554, but some experts have found their clinical data to be less impressive than their marketing acumen.

Differentiating bold claims from promising avenues can end up becoming a full-time job, according to Susan Holmes, a hair loss expert at the British Association of Dermatologists.

'If we're asked to comment on these things,' she says, 'it's weeks of research to go through the literature. What we want to see is good evidence in a research paper, research that's been done properly, that's been peer-reviewed. A lot of the literature is in small journals where it's difficult to know who's reviewed it and what process it's gone through. There's a lot of research being done, and a lot of avenues look interesting, but it's a question of whether they come through all the rigorous tests to become an effective treatment. Hair is just a difficult thing to make grow.'

§ § §

Hair is just a difficult thing to make grow. This should be the disclaimer on every medicine, every 'natural' supplement, every website that claims to offer treatments for hair loss, but instead they choose to sell false hope. 'A cure is always five years away,' says Stevenson, with a laugh. 'In five years, it will be five years away – in 10 years it'll be five years away. It's the Holy Grail. But I think cloning will be a massive industry.'

Successful cloning of hair follicles could give transplant patients a more plentiful source of hair; currently that source is restricted to their very valuable but very limited donor area. 'Hair cloning, neogenesis, induction, it's all the same thing,' says Claire Higgins, a lecturer in the Department of Bioengineering at Imperial College London, 'but it's really hard to do.' In a 2013 paper, five scientists including Higgins described how they managed to initiate neogenesis in human skin. 'We took human skin and stuck it on a mouse,' she says. 'It almost acts like an oven, to cook the tissue, but the hairs were tiny. We were inducing hair in an epidermis that wasn't very receptive. We initiated it, but something inhibited it. I think it will work, long-term, but in our lab we're trying to map out the genetic changes that are occurring before trying induction again.'

The work done in Higgins's lab isn't driven by a search for a cure for hair loss, but hair happens to be a convenient, accessible model for her to work with. 'We think that [during the miniaturisation that causes male pattern hair loss] cells are migrating away from the hair into the surrounding skin,' she says. 'The reverse process of that occurs during hair development. Cells migrate together, you get a cluster of cells that's about double the density of the surrounding cells, which goes on to become the dermal papilla. If I can use the hair to understand this process, how the hair can reprogramme the epidermis to change its identity, I think these are basic biological questions that can apply to very different systems.'

§ § §

Hair regrowth may not be the specific focus of her research, but Higgins believes that the psychological issues caused by male pattern hair loss make that work more valuable. 'People won't leave the house,' she says. 'It's not life-threatening, but it is life-changing.' It certainly changed Jay Patel's life, although he now feels that he's pulled things around. 'I'm 37 now,' he says. 'I was 23 when I started losing my hair, so it took me a long time. I wasn't a confident person. I wouldn't have opened a business like this, because I'm not that sort of person. I'm the person who sits in the corner, in the background. I went through a lot of hardship to get where I am.'

Susan Holmes mourns the lack of psychological support available on the NHS. 'What little there is is hugely over-subscribed,' she says. 'There are many people with many different disorders who require the input of a clinical psychologist. We know we can't cure hair loss, we know that what men need is help to come to terms with it.' But given our poor track record, is there any likelihood that balding men like me will ever relish our slow shedding of hair?

Milan Stolicny hopes so. His website, baldattraction.com, offers a joyous, upbeat appreciation of balding heads, and beseeches men with receding hairlines to reclaim their baldness and relish the new perspective it gives them. 'Bald Is Very Attractive!' he says. 'Bald can wildly succeed in this world! It's time to kick ass and wildly succeed as a bald man!' Stolicny offers no quackery, no remedies, potions or balsams – just enthusiasm. 'The true baldness cure,' he says, 'is to become attractive bald man!' In other words, believe in yourself. It looks so easy, written down. If balding men were able to do such a thing, a huge industry would collapse overnight. But that industry knows only too well that Stolicny's solution, while beautifully simple, is perhaps the hardest of all to achieve.

This story was first published on 4 September 2016 by Wellcome on mosaicscience.com

How a bee sting saved a woman's life

■ Christie Wilcox

'I moved to California to die.'

Ellie Lobel was 27 when she was bitten by a tick and contracted Lyme disease. And she was not yet 45 when she decided to give up fighting for survival.

Caused by corkscrew-shaped bacteria called Borrelia burgdorferi, which enter the body through the bite of a tick, Lyme disease is diagnosed in around 300,000 people every year in the United States. It kills almost none of these people, and is by and large curable – if caught in time. If doctors correctly identify the cause of the illness early on, antibiotics can wipe out the bacteria quickly before they spread through the heart, joints and nervous system.

But back in the spring of 1996, Ellie didn't know to look for the characteristic bull's-eye rash when she was bitten – she thought it was just a weird spider bite. Then came three months with flu-like symptoms and horrible pains that moved around the body. Ellie was a fit, active woman with three kids, but her body did not know how to handle this new invader. She was incapacitated. 'It was all I could do to get my head up off the pillow,' Ellie remembers.

Her first doctor told her it was just a virus, and it would run its course. So did the next. As time wore on, Ellie went to doctor after doctor, each giving her a different diagnosis. Multiple sclerosis. Lupus. Rheumatoid arthritis. Fibromyalgia. None of them realised she was infected with Borrelia until more than a year after she contracted the disease – and by then, it was far too late. Lyme bacteria are exceptionally good at adapting, with some evidence that they may be capable of dodging both the immune system and the arsenal of antibiotics currently available. Borrelia are able to live all over the body, including the brain, leading to neurological symptoms. And even with antibiotic treatment, 10–20 per cent of patients don't get better right away. There are testimonies of symptoms persisting – sometimes even resurfacing decades after the initial infection – though the exact cause of such post-treatment Lyme disease syndrome is a topic of debate among Lyme scientists.

'I just kept doing this treatment and that treatment,' says Ellie. Her condition was constantly worsening. She describes being stuck in bed or a wheelchair, not being able to think clearly, feeling like she'd lost her short-term memory and not feeling 'smart' anymore. Ellie kept fighting, with every antibiotic, every pharmaceutical, every holistic treatment she could find. 'With some things I would get better for a little while, and then I would just relapse right back into this horrible Lyme nightmare. And with every relapse it got worse.'

After fifteen years, she gave up.

'Nothing was working any more, and nobody had any answers for me,' she says. 'Doctors couldn't help me. I was spending all this cash and was going broke, and when I got my last test results back and all my counts were just horrible, I knew right then and there that this was the end.'

'I had outlived so many other people already,' she says, having lost friends from Lyme support groups, including some who just couldn't take the suffering any more. 'I didn't care if I was going to see my next birthday. It's just enough. I was ready to call it a life and be done with it.'

So she packed up everything and moved to California to die. And she almost did.

Less than a week after moving, Ellie was attacked by a swarm of Africanised bees.

§ § §

Ellie was in California for three days before her attack. 'I wanted to get some fresh air and feel the sun on my face and hear the birds sing. I knew that I was going to die in the next three months or four months. Just laying there in bed all crumpled up... It was kind of depressing.'

At this point, Ellie was struggling to stand on her own. She had a caregiver on hand to help her shuffle along the rural roads by her place in Wildomar, the place where she had chosen to die.

She was just standing near a broken wall and a tree when the first bee appeared, she remembers, 'just hitting me in the head'. 'All of a sudden – boom! – bees everywhere.'

Her caregiver ran. But Ellie couldn't run – she couldn't even walk. 'They were in my hair, in my head, all I heard was this crazy buzzing in my ears. I thought: wow, this is it. I'm just going to die right here.'

Ellie, like 1–7 per cent of the world's population, is severely allergic to bees. When she was two, a sting put her into anaphylaxis, a severe reaction of the body's immune system that can include swelling, nausea and narrowing of the airways. She nearly died. She stopped breathing and had to be revived by defibrillation. Her mother drilled a fear of bees into her to ensure she never ended up in the same dire situation again. So when the bees descended, Ellie was sure that this was the end, a few months earlier than expected.

Bees – and some other species in the order Hymenoptera, such as ants and wasps – are armed with a potent sting that many of us are all too aware of. This is their venom, and it's a mixture of many compounds. Perhaps the most important is a tiny 26-amino-acid peptide called melittin, which constitutes more than half of the venom of honey bees and is found in a number of other bees and wasps. This little compound is responsible for the burning pain associated with bee stings. It tricks our bodies into thinking that they are quite literally on fire.

When we experience high temperatures, our cells release inflammatory compounds that activate a special kind of channel, TRPV1, in sensory neurons. This ultimately causes the neurons to send a signal to the brain that we're burning. Melittin subversively makes TRPV1 channels open by activating other enzymes that act just like those inflammatory compounds.

Jellyfish and other creatures also possess TRPV1-activating compounds in their venoms. The endpoint is the same: intense, burning pain.

'I could feel the first five or ten or fifteen but after that... All you hear is this overwhelming buzzing, and you feel them hitting your head, hitting your face, hitting your neck,' says Ellie. 'I just went limp. I put my hands up and covered my face because I didn't want them stinging me in the eyes... The next thing I know, the bees are gone.'

When the bees finally dissipated, her caregiver tried to take her to the hospital, but Ellie refused to go. 'This is God's way of putting me out of my misery even sooner,' she told him. 'I'm just going to accept this.'

'I locked myself in my room and told him to come collect the body tomorrow.'

But Ellie didn't die. Not that day, and not three to four months later. 'I just can't believe that was three years ago, and I just can't believe where I am now,' she tells me. 'I had all my blood work done. Everything. We tested everything. I'm so healthy.'

She believes the bees, and their venom, saved her life.

§ § §

The idea that the same venom toxins that cause harm may also be used to heal is not new. Bee venom has been used as a treatment in East Asia since at least the second century BCE. In Chinese traditional medicine, scorpion venom is recognised as a powerful medicine, used to treat everything from eczema to epilepsy. Mithradates VI of Pontus, a formidable enemy of Rome (and also an infamous toxinologist), was said to have been saved from a potentially fatal wound on the battlefield by using steppe viper venom to stop the bleeding.

'Over millions of years, these little chemical engineers have developed a diversity of molecules that target different parts of our nervous system,' says Ken Winkel, Director of the Australian Venom Research Unit at the University of Melbourne. 'This idea of applying these potent nerve toxins to somehow interrupt a nervous disease has been there for a long time. But we haven't known enough to safely and effectively do that.'

Despite the wealth of history, the practical application of venoms in modern therapeutics has been minimal. That is, until the past ten years or so, according to Glenn King at the University of Queensland in Brisbane, Australia. In 1997, when Ellie was bouncing around from doctor to doctor, King was teasing apart the components of the venom from the Australian funnel-web, a deadly spider. He's now at the forefront of venom drug discovery.

King's group was the first to put funnel-web venom through a separation method called high-performance liquid chromatography (HPLC), which can separate out different components in a mixture based on properties like size or charge. 'I was just blown away,' he says. 'This is an absolute pharmacological goldmine that nobody's really looked at. Clearly hundreds and hundreds of different peptides.'

Over the course of the 20th century, suggested venom treatments for a range of diseases have appeared in scientific and medical literature. Venoms have been shown to fight cancer, kill bacteria, and even serve as potent painkillers – though many have only gone as far as animal tests. At the time of writing, just six had been approved by the US Food and Drug Administration for medical use (one other – Baltrodibin, adapted from the venom of the Lancehead snake – is not FDA approved, but is available outside the US for treatment of bleeding during operations).

The more we learn about the venoms that cause such awful damage, the more we realise, medically speaking, how useful they can be. Like the melittin in bee venom.

Melittin does not only cause pain. In the right doses, it punches holes in cells' protective membranes, causing the cells to explode. At low doses, melittin associates with the membranes, activating lipid-cutting enzymes that mimic the inflammation caused by heat. But at higher concentrations, and under the right conditions, melittin molecules group together into rings creating large pores in membranes, weakening a cell's protective barrier and causing the entire cell to swell and pop like a balloon.

Because of this, melittin is a potent antimicrobial, fighting off a variety of bacteria and fungi with ease. And scientists are hoping to capitalise on this action to fight diseases like HIV, cancer, arthritis and multiple sclerosis.

For example, researchers at the Washington University School of Medicine in St Louis, Missouri, have found that melittin can tear open HIV's protective cell membrane without harming human cells. This envelope-busting method also stops the virus from having a chance to evolve resistance. 'We are attacking an inherent physical property of HIV,' Joshua L Hood, the lead author of the study, said in a press statement. 'Theoretically, there isn't any way for the virus to adapt to that. The virus has to have a protective coat.' Initially envisioned as a prophylactic vaginal gel, the hope is that melittin-loaded nanoparticles could someday be injected into the bloodstream, clearing the infection.

§ § §

Ellie is the first to admit that her tale sounds a little tall. 'If someone were to have come to me and say, 'Hey, I'll sting you with some bees, and you'll get better', I would have said, 'Absolutely not! You're crazy in your head!''. But she has no doubts now.

After the attack, Ellie watched the clock, waiting for anaphylaxis to set in, but it didn't. Instead, three hours later, her body was racked with pains. A scientist by education before Lyme took its toll, Ellie thinks that these weren't a part of an allergic response, but instead indicated a Jarisch–Herxheimer reaction – her body was being flooded with toxins from dying bacteria. The same kind of thing can happen when a person is cured from a bad case of syphilis. A theory is that certain bacterial species go down swinging, releasing nasty compounds that cause fever, rash and other symptoms.

For three days, she was in pain. Then, she wasn't.

'I had been living in this... I call it a brown-out because it's like you're walking around in a half-coma all the time with the inflammation of your brain from the Lyme. My brain just came right out of that fog. I thought: I can actually think clearly for the first time in years.'

With a now-clear head, Ellie started wondering what had happened. So she did what anyone else would do: Google it. Disappointingly, her searches turned up very little. But she did find one small 1997 study by scientists at the Rocky Mountain

Laboratories in Montana, who'd found that melittin killed Borrelia. Exposing cell cultures to purified melittin, they reported that the compound completely inhibited Borrelia growth. When they looked more closely, they saw that shortly after melittin was added, the bacteria were effectively paralysed, unable to move as their outer membranes were under attack. Soon after, those membranes began to fall apart, killing the bacteria.

Convinced by her experience and the limited research she found, Ellie decided to try apitherapy, the therapeutic use of materials derived from bees.

Her bees live in a 'bee condo' in her apartment. She doesn't raise them herself; instead, she mail orders, receiving a package once a week. To perform the apitherapy, she uses tweezers to grab a bee and press it gently where she wants to be stung. 'Sometimes I have to tap them on the tush a little bit,' she says, 'but they're usually pretty willing to sting you.'

She started on a regimen of ten stings a day, three days a week: Monday, Wednesday, Friday. Three years and several thousand stings later, Ellie seems to have recovered miraculously. Slowly, she has reduced the number of stings and their frequency – just three stings in the past eight months, she tells me (and one of those she tried in response to swelling from a broken bone, rather than Lyme-related symptoms). She keeps the bees around just in case, but for the past year before I talked to her, she'd mostly done just fine without them.

§ § §

Modern science has slowly begun to take apart venoms piece by piece to understand how they do the things they do, both terrible and tremendous. We now know that most venoms are complex cocktails of compounds, with dozens to hundreds of different proteins, peptides and other molecules to be found in every one. The cocktails vary between species and can even vary within them, by age, location or diet. Each compound has a different task that allows the venom to work with maximum efficiency – many parts moving together to immobilise, induce pain, or do whatever it is that the animal needs its venom for.

The fact that venoms are mixtures of specifically targeted toxins rather than single toxins is exactly what makes them such rich sources of potential drugs – that's all a drug is, really, a compound that has a desired effect on our bodies. The more specific the drug's action, the better, as that means fewer side-effects.

'It was in the 2000s that people started saying well, actually, [venoms] are really complex molecular libraries, and we should start screening them against specific therapeutic targets as a source of drugs,' says King.

Of the seven venom-derived pharmaceuticals on the international market, the most successful, captopril, was derived from a peptide found in the venom of the Brazilian viper (Bothrops jararaca). This venom has been known for centuries for its potent blood-thinning ability – one tribe are said to have coated their arrow tips in it to inflict maximum damage – and the drug has made its parent company more than a billion dollars and become a common treatment for hypertension.

Bryan Fry, a colleague of Glenn King's at the University of Queensland and one of the world's most prolific venom researchers, says the captopril family and its derivatives still command a market worth billions of dollars a year. Not bad for something developed in 1970s. 'It's not only been one of the top twenty drugs of all time,' he says, 'it's been one of the most persistent outside of maybe aspirin.'

And it's not just captopril. Fry points to exenatide, a molecule found in the venom of a lizard, the gila monster, and the newest venom-derived pharmaceutical on the US market. Known by the brand name Byetta, this has the potential to treat type 2 diabetes, stimulating the body to release insulin and slow the overproduction of sugar, helping reverse the hormonal changes caused by the disease.

Rare cases like Ellie's are a reminder of the potent potential of venoms. But turning folk knowledge into pharmaceuticals can be a long and arduous process. 'It could take as long as ten years from the time you find it and patent it,' says King. 'And for every one that you get through, ten fail.'

§§§

Since the 1997 study, no one had looked further into bee venom as a potential cure for Lyme disease, until Ellie.

Ellie now runs a business selling bee-derived beauty products called BeeVinity, inspired after, she says, noticing how good her skin looked as she underwent apitherapy. 'I thought, "Well, people aren't going to want to get stung with bees just to look good."'

Ellie has partnered with a bee farm that uses a special electrified glass plate to extract venom. As the bees walk across the plate on the way to and from their hive, harmless currents stimulate the bees to release venom from their abdomens, leaving teeny little droplets on the glass, which are later collected. Ellie says it takes 10,000 bees crossing that plate to get 1 gram of venom (other sources, such as the Food and Agriculture Organization of the UN, quote 1 million stings per gram of venom), but 'those bees are not harmed'.

For her, it is more than just a way to make a living: it's 'an amazing blessing'. Proceeds from her creams and other products support bee preservation initiatives, as well as Lyme disease research. In addition, she sends some of the venom she purchases – which, due to the cost of the no-harm extraction method she uses, she says is 'more expensive than gold' – to Eva Sapi, Associate Professor of Biology and Environmental Science at the University of New Haven, who studies Lyme disease.

Sapi's research into the venom's effects on Lyme bacteria is ongoing and as yet unpublished, though she told me the results from preliminary work done by one of her students look 'very promising'. Borrelia bacteria can shift between different forms in the body, which is part of what makes them so hard to kill. Sapi has found that other antibiotics don't actually kill the bacteria but just push them into another form that is more dormant. As soon as you stop the antibiotics, the Borrelia bounce back. Her lab is testing different bee venoms on all forms of the bacteria, and so far, the melittin venom seems effective.

The next step is to test whether melittin alone is responsible, or whether there are other important venom components. 'We also want to see, using high-resolution images, what exactly happens when bee venom hits Borrelia,' Sapi told me.

She stresses that much more data is needed before any clinical use can be considered. 'Before jumping into the human studies, I would like to see some animal studies,' she says. 'It's still a venom.' And they still don't really know why the venom works for Ellie, not least because the exact cause of post-treatment Lyme disease symptoms remains unknown. 'Is it effective for her because it's killing Borrelia, or is it effective because it stimulates the immune system?' asks Sapi. It's still a mystery.

There's a long way to go for bee venom and melittin. And it takes a lot of work – and money – to turn a discovery into a safe, working medicine. But labs like King's are starting to tap the pharmaceutical potential that lies in the full diversity of venomous species. And King, for one, believes that scientists are entering a new era of drug discovery.

In the past, venoms have been investigated because of their known effects on humans. Such investigations required both knowledge of the venom's clinical effects and large volumes of venom, so until now only large species, like snakes, with easily extracted venoms have been studied in any depth. But that's changing. Technological advances allow for more efficient venom extraction as well as new ways to study smaller amounts of venom. The preliminary tests for pharmaceuticals can now start with nothing more than a genetic sequence. 'We can now genomically look at the toxins in these animals without having to actually even purify the venom,' says King, 'and that changes everything.' Ken Winkel thinks venomous animals will be excellent drug resources for devastating neurological diseases, as so many of their venoms target our nervous system. 'We really don't have great drugs in this area,' he says, 'and we have these little factories that have a plethora of compounds...'

No one knows exactly how many venomous species there are on this planet. There are venomous jellyfish, venomous snails, venomous insects, even venomous primates. With that, however, comes a race against time of our own making. Species are going extinct every year, and up to a third may go extinct from climate change alone.

'When people ask me what's the best way to convince people to preserve nature, your weakest argument is to talk about how beautiful and wonderful it is,' says Bryan Fry. Instead, he says, we need to emphasise the untapped potential that these species represent. 'It's a resource, it's money. So conservation through commercialisation is really the only sane approach.'

Ellie couldn't agree more. 'We need to do a lot more research on these venoms,' she tells me emphatically, 'and really take a look at what's in nature that's going to help us.'

This story was first published on 24 March 2015
by Wellcome on mosaicscience.com

The global trend for 'kangaroo' babies

■ Lena Corner

Carmela Torres was 18 when she became pregnant for the first time. It was 1987 and she and her now-husband, Pablo Hernandez, were two idealistic young Colombians born in the coastal region of Montería who moved to the capital, Bogotá, in search of freedom and a better life. When Torres told her father she was expecting, so angered was he by the thought of his daughter having a child out of wedlock that they didn't speak to each other for years.

Torres remained undaunted. Her pregnancy was trouble-free and she had a new life in Bogotá to get on with. But one December afternoon, suddenly, out of nowhere, her body began to convulse with sharp contractions. It was more than two months before her due date. She called Hernandez and together they rushed to the Instituto Materno Infantil (Mother and Child Hospital) in eastern Bogotá.

Not long after arriving she gave birth naturally to a baby boy weighing just 1,650 grams (3 lb 10 oz).

Before she had a chance to hold him, her baby was whisked off to a neonatal intensive care unit. Torres was simply told to get dressed and go home. 'I didn't even get to touch him,' she says. 'They said I could come back and see him but the visiting times were very restricted – just a couple of hours a day. When I did visit I was allowed to look but not touch.'

On the third day she was at home preparing for her next visit when the phone rang. 'It was the hospital,' she says. 'They called to say my baby was dead. They didn't tell me the cause of death or give me any diagnosis. Just that he was dead. I hadn't even named him yet.'

Torres was traumatised. She became gripped with an acute feeling of isolation and started spiralling into depression. She knew she needed to do something to pull herself back so she enrolled on a teacher-training programme and immersed herself in her studies. 'It gave me something to focus on,' she says. 'It saved me.'

A decade passed before Torres was ready to become pregnant again. This time round it was different. By now she was married to Hernandez and well settled in Bogotá. Her father had even started talking to her again. She was so excited about giving birth that, a couple of months before her due date, she decided to throw a big baby shower. But on the day of the party, familiar, severe contractions ripped through her body, stopping her in her tracks. She smiled, told no one and pretended it wasn't happening. By evening, once all the guests were gone, she could hide it no longer. She told Hernandez who again rushed her straight to the Instituto Materno Infantil.

'When we got there the doctor was furious with me for not coming earlier. He said I was ready to give birth,' says Torres. 'I was petrified, I didn't want another premature baby. I was taken to the exact same ward as where I had my baby which died. Memories came flooding back. I was extremely stressed.'

At one o'clock the next morning Torres gave birth to another boy. She named him immediately, calling him Julian. He weighed almost the same as her firstborn and just like then, he was whisked straight into intensive care. History, it seemed, was repeating itself.

'I spent a very frightening night panicking that I was about to lose another baby,' she says. 'But the next morning a doctor came to see me. She told me about a thing called Kangaroo Mother Care – how I could act as a human incubator and carry my own baby and take it home with me. It was a ray of light at the end of the tunnel. Anything rather than leave my baby there.'

That day Torres was taught how to hold her baby under her clothing, upright between her breasts with his airways clear. She was taught how even the finest layer of fabric between her and her baby wasn't allowed – it had to be continuous and direct skin-to-skin contact. She was taught how to breastfeed, how to sleep on her back propped up by cushions, and strictly never to bathe him as this would waste his precious energy. Remarkably, the very next afternoon, with her tiny baby strapped to her chest under a blanket, Torres walked out of hospital.

'Julian was very small and fragile but I was much happier taking him home with me than leaving him there, where my other baby had died,' she says. 'Feeding him wasn't easy, but I had a lot of help. At first I had to go back every day for follow-up appointments and I was given a cellphone number that I could call any time I needed. We had to go back in when Julian got an infection on his umbilical cord and for phototherapy when he got jaundice, but in all I carried him for a month 24 hours a day, sharing shifts with my husband, until he hit his target weight of 2,500 g. Once he'd reached that we didn't have to do it any more and finally he got his first bath.'

§ § §

Kangaroo Mother Care (KMC) is the brainchild of Colombian paediatrician Edgar Rey, who introduced it to the Instituto Materno Infantil in 1978. It was an idea born out of desperation. The institute served the city's poorest – those who lived crammed in the rickety makeshift dwellings in the foothills of the surrounding mountains. At the time this was the biggest neonatal unit in all of Colombia, responsible for delivering 30,000 babies a year. Overcrowding was so bad that three babies would have to share an incubator at a time and the rate of cross-infection was high. Death rates were spiralling and so too was the level of abandonment as young, impoverished mothers, who never even got to touch their babies, found it easier just to take off.

Scouting around for a solution to these problems, Rey happened upon a paper on the physiology of the kangaroo. It mentioned how at birth kangaroos are bald and roughly the size of a peanut – very immature, just like a human pre-term baby. Once in its mother's pouch the kangaroo receives thermal regulation from the direct skin-to-skin contact afforded by its lack of hair. It then latches onto its mother's nipple, where it remains until it has grown to roughly a quarter of its mother's weight, when finally it is ready to emerge into the world.

This struck a chord with Rey. He went back to the institute and decided to test it out. He trained mothers of premature babies to carry them just as kangaroos do. Working alongside his colleague Hector Martinez, he taught them the importance of breastfeeding and discharged them just as soon as their babies were able. The results were remarkable. Death rates and infection levels dropped immediately. Overcrowding was reduced because hospital stays were much shorter, incubators were freed up, and the number of abandoned babies fell.

§ § §

It's 8am and already the shiny new KMC unit at the San Ignacio University Hospital in downtown Bogotá is packed to the rafters. Rows and rows of women, and a surprisingly high number of men, too, squeeze together – a sea of colourful knitted hats and chunky coats, protection against the city's unpredictable cycle of hail, rain and heat. They sit on narrow pews in the centre of the room, with the tiniest little heads peeping skyward on their chests. It's warm, buzzy and a million miles away from the sterile atmosphere of a typical neonatal intensive care unit.

Many seem to have settled in for the day – one woman has her knitting out and another has her extended family in tow. Five paediatricians stand in a row behind a long, high bench examining baby after baby, testing their responses and bending their limbs this way and that. On an average day they will see more than 100. For a room full of newborns, it's oddly peaceful. Not one of them is screaming.

'Traditional units are closed and have very restrictive visiting hours,' says Nathalie Charpak, the formidable French paediatrician who now heads the unit and lives just a short walk away. 'An important element of KMC is that the unit is open and parents have access so they can sit with their infants, connect with each other and gain confidence seeing others with very small babies doing the same thing. Evidence shows there is less infection when units are open like this because the parents are checking to ensure the health professionals are washing their hands.'

In one corner intensive breastfeeding sessions are taking place. Eleanora Rodrigez, a raven-haired 26-year old from northern Bogotá, had just returned from a walk in the park when her waters unexpectedly broke. She gave birth to twins Henry (1,700 g) and Joaquim (1,450 g) at 32 weeks. Her slightly jittery husband hovers about trying to second-guess their every need. Today, Rodrigez is being taught how to give surprisingly tough massages across her babies' heads, foreheads, upper lips and chins to stimulate their sucking motion. Joaquim, in particular, keeps nodding off.

'It's really hard,' says Rodrigez, struggling to untangle both her babies from their oxygen tubes. 'They are feeding every two hours. They have to gain 15 g per kg every day, the same growth as they would have had in the womb. If this is happening we know things are OK. I'm just waiting till they hit the magic 2,500 g.'

In a side room, a clinical psychologist is doing a session with a small group of mostly nervous-looking teenage mums. One, wearing grubby grey tracksuit bottoms and what appears to be her boyfriend's football shirt, looks barely into her teens. Her baby is so tiny you might be scared to even touch it, yet she wanders around with it dangling from her arm with an insouciance that only youth can bring. All the babies here are born at less than 37 weeks or weigh below 2,500 g. Yesterday a bubbly 11-year-old girl popped in with her mother. She had been a kangaroo baby herself, born here at 29 weeks and weighing just 500g.

§ § §

Charpak moved from France to Bogotá in 1986, after falling in love with a Colombian university professor. She wound up at the Instituto Materno Infantil working alongside Rey and Martinez. Charpak couldn't believe the results she was witnessing. She understood immediately the need for rigorous scientific studies to prove to the world they were on to something very special.

In 1989, she did a study on a sample of babies from two of the very poorest hospitals in the city. She proved scientifically that KMC was safe – even the smallest premature babies weren't dying if you took them out of the incubator. In 1994, with funding from a Swiss NGO, she did a much larger randomised trial, which proved conclusively that not only were babies dying less, but breastfeeding rates were up, hospital stays were shorter and infection was down. Her findings were published in Pediatrics journal in 1997.

Charpak now lives with her husband and two sons (a third is studying in St Petersburg) in a large apartment at the top of a redbrick tower on Seventh Street, a broad thoroughfare that sweeps regally through the city. Her balcony looks out towards the astonishing orange sunsets over the city on one side and the lush mountain of Monserrate on the other. She's in her 60s now and has lived in Bogotá for 30 years. Every day she makes the short walk over the uneven pavements, through the famously perilous traffic, to her unit at the San Ignacio. Her father, Georges Charpak, won the Nobel Prize in Physics in 1992, for the invention and development of particle detectors. For Nathalie, KMC has become her life's work. On top of her clinical work, she is director of an NGO that researches and promotes KMC, the Fundación Canguro – the Kangaroo Foundation.

Employed alongside her at the Foundation is a young, glamorous and smart sidekick, Julieta Villegas, who is clearly being groomed to take on the KMC mantle whenever Charpak is ready to step aside, which isn't any time soon. 'I'm Nathalie's replacement,' is Villegas's brazen introduction. The pair are an undeniable force, driven by an unwavering belief in KMC. It's hard not to share their enthusiasm. There are now more than 1,600 studies which show KMC does so much more than just help a baby put on weight. Research shows, for example, a kangaroo baby bonds better with its parents than pre-term babies given conventional care. Its heart and respiratory rates improve better. It is more equipped to self-regulate and so is calmer and better able to sleep. Kangaroo mothers, too, perhaps given a sense of purpose after the guilt they often feel about giving birth prematurely, experience less post-natal depression. And, most remarkably, tests done on kangaroo babies at 12 months old show they have higher IQs and better cognitive development than those given conventional care.

'It is clear KMC is about much more than just saving the baby's life,' says Charpak. 'It is about allowing the baby to thrive and giving it the best possible quality of life. I have fought all my life to show that KMC has nothing to do with comfort or massage or anything fluffy like that. It is difficult to do and each baby is carefully followed up every six weeks for the first year, but the benefits are extraordinary.'

Thanks, in large part, to Charpak's persistence in pushing all the research under the noses of the health ministry, KMC is now enshrined in Colombian law. All women with premature or low-weight babies will be sent as a matter of course to their nearest centre. We drive out to Tunja in the eastern Andes to see the unit there. Today, flash floods have turned roads into streaming rivers of grey mud, and alarming-looking stray dogs roam wildly. It's much poorer here and there's a palpable edge of desperation. Most of the locals are farmers who make a living growing potatoes and corn. The KMC unit at the San Rafael Hospital is run by local paediatrician Jenny Lizarazo Medina. She tells me that about 40 per cent of the women here have low-birthweight babies not because they are born prematurely but simply because when their mothers were pregnant, they went hungry.

Maria 24, is one. She carried her daughter Natalia to term but the baby weighed just 2,170 g at birth. Wearing a baggy turquoise tracksuit, eyes shining with exhaustion, Maria arrives for her daily check dragging a bulky metal oxygen tank behind her, just as she would a shopping trolley. Tunja is 3,000 m above sea level, one of the highest cities in Colombia, so a lot of these babies need extra oxygen.

It's remote here and distance to the hospital is an issue. Maria has temporarily moved in with an uncle who lives nearby while her husband remains at their home in Cómbita, further north, working in a recycling factory. This means Maria is doing everything alone, including carrying her baby and dragging the oxygen tank up a steep mountain every day to get back to her room. In the very early days Natalia found it hard to latch on so Maria sat for hours, day and night, feeding her breast milk out of a tiny cup. 'Each day my baby changes and is now making good progress,' says Maria. 'And every day I come here they say good things about her. It gives me confidence.'

§ § §

One of the very first countries to investigate what was going on in Bogotá was Venezuela. In 1994 a small team, from a hospital just like the Instituto Materno Infantil, came to witness KMC for themselves. Others, mainly from low-income countries, came too. Brazil in 1995, Ethiopia in 1996, followed shortly by Madagascar, India, Cameroon and many more. Charpak used to put them up in her own home and give them 15 days of intensive training. Nowadays they stay in the Kangaroo Foundation headquarters in Bogotá. Then they return home to run their own KMC programmes.

Many of these are very successful. In Malawi, which has the highest rate of premature births in the world (181 babies out of 1,000), there is now a KMC centre in every district. Over the ten years to 2015, the number of babies dying before their first birthday fell from 72 out of 1,000 to 43. 'I have seen a significant drop in mortality,' says Indira, a midwife at Zomba Central Hospital in southern Malawi. 'It has also helped reduce congestion in the ward as babies are cared for at home. And it has helped reduce costs, because electricity is being saved as the mother is a perfect heat source for the baby.'

In Cameroon, a country Charpak has been back and forth to many times, mortality of premature babies has dropped by around 30 per cent. According to a 2016 analysis by Cochrane, a global network of researchers, studies of KMC have found it to reduce mortality in pre-term newborns by 33 per cent. The World Health Organization has estimated that KMC has the potential to save as many as 450,000 lives a year.

Resistance, however, has come from where you might least expect it. For some health professionals, nurses and even paediatricians, Charpak tells me, it can be difficult to accept that care by mothers is better than anything they can offer themselves, especially if they have fought hard to bring shiny rows of incubators to their hospitals. There is also the prevailing idea that things are done better in Westernised countries.

While the idea for KMC may have been a product of necessity, Charpak and Villegas are now locked in a constant struggle to convince the world that it isn't just the poor woman's option. 'It's not a cheap alternative. It's not something just to be done in poor countries,' says Charpak. 'There is a cost to it. It's a proper neonatal care with advantages that are clinically proven.'

Undeniably, though, it is cheaper. The estimated cost of neonatal care for premature babies in the United States is $3,000–5,000 a day. In contrast, in low-income countries a KMC programme can cost as little as $4.60 a day.

One early visitor to Bogotá was Susan Ludington from Case Western Reserve University in Cleveland, Ohio. She went to visit Charpak in 1988 after seeing a short video of KMC being practised in Bogotá.

'I brought a team of researchers and moved into the Instituto Materno Infantil and I was quite blown away by what I saw,' says Ludington. 'The babies were so quiet and so calm. They were sleeping deeply and then they would wake up and suck with vigour. Our pre-term babies were upset all the time and not sleeping well. They were all in incubators. The mothers came to visit but they certainly weren't holding them. I thought it was really impressive. I had many, many questions.'

Ludington returned home and tried to get some interest in doing research on KMC in the US. 'I went to 18 different hospitals in the LA area. All of them turned me down. They said you should study this first on apes, and why would we want to put a premature baby on a mother's smelly, sweaty breast? Or, the baby will get cold and then we will be sued. You have to remember in the States as recently as the 1970s mothers were not even allowed into the intensive care unit to see a premature baby until the 21st day of its life. By 1988, when I am proposing skin-to-skin, the mothers are allowed in but are not allowed to take the babies out of the incubator, only to put a hand inside on the baby's thigh.'

Eventually Ludington found support from the head of neonatology at the Hollywood Presbyterian Hospital. He agreed to let her do a study – the first ever in the US. 'We were trying to determine if it was safe and we found yes, it was safe, better than safe,' says Ludington. 'We now know the best protection from infection [for the baby] is to be colonised by its mother's bacteria. We also know the best thing for its brain development is skin-to-skin, the best way to maintain blood sugar levels so it doesn't get hyperglycaemic is skin-to-skin. And what we didn't know in 1988 was that there are a whole set of nerves on the baby's chest and on the mother's chest that only get stimulated by skin-to-skin contact, which send oxytocin messages to the baby's brain.'

According to the map on the wall in the offices of the Kangaroo Foundation in Bogotá, there are now KMC centres in almost 70 countries across the world, including Australia, Spain and France. This simple yet smart idea, which originated amid grinding poverty, is now spreading to the very richest parts of our globe. In parts of Scandinavia it is offered as a matter of course. The University Children's Hospital in Uppsala, Sweden, is leading the way and is now doing skin-to-skin with mothers of babies born at just 25 weeks.

So why haven't we all ditched our incubators to take up KMC? 'It's just not that simple,' says Angela Huertas, consultant neonatologist at University College Hospital in London. 'In Uppsala the mother is admitted to hospital as soon as the baby is admitted. There, they have two years maternity and one year paternity leave so they have the support to do that. Even if we desperately wanted to do KMC here in UCLH we wouldn't be able to do it. We don't have the infrastructure for mothers to stay, nor the nursing staff for support. Families here are different from those in Colombia or Scandinavia. People have to work, and who is going to look after their other children? It's not just the parent's wishes or the baby's needs, you've got to have the political will to make changes as big as this.'

Also, there are times when Huertas believes babies are actually better off in incubators. 'Very premature babies have skin so thin it's almost transparent – you can see their veins through it,' she says. 'These babies are clearly better off inside an incubator as they need a humid environment until their skin can thicken.'

She thinks we need to find a balance. 'There are many neonatal units in London that won't allow the parents in more than one hour a day,' she says. 'They say it's quiet time and send the parents away so the babies get looked after by the nurses. There is an awful lot we can do about that.'

§ § §

Last November a group including Charpak, Villegas and Ludington came together for a meeting in Trieste, Italy. It was a seminal moment, marking the 20th anniversary of the first ever global KMC meeting – now with a much bigger and vocal group. Charpak and Villegas unveiled the most ambitious study yet into KMC. They had decided to see if they could track down the 716 families who took part in Charpak's original 1994 study and measure the effects of KMC at 20 years to see if any of the benefits persisted. They launched a nationwide search, advertising on the radio, on TV and in the press to find the original kangaroo mothers. 441 of them came forward.

One of the first to volunteer was Carmela Torres. It turns out that seven years after giving birth to Julian she had had a third premature baby. This one, called Pablo, was born at 33 weeks and weighed 1,600 g – even less than his brothers. Torres remembered Charpak well from Julian's birth. 'She treated him like family. When he had to go back in to hospital with an umbilical cord infection I'd go in early every morning and last thing at night to do kangaroo care. If ever I was late I'd find that Dr Charpak had picked Julian up and was carrying him in the kangaroo position herself. Doing KMC the second time round was completely different. I was confident. I knew exactly what I was doing. Pablo gained weight much faster than Julian.'

So Julian, now aged 22, along with all these other original kangaroo babies, was subjected to a series of rigorous checks including MRIs, neuroimaging, blood tests, psychosocial tests and physical evaluations. Each was measured for self-esteem, depression, hyperactivity, aggressiveness and more. So were the grown-up babies from the original control group, who had received traditional care. The full results were published in Pediatrics journal at the end of last year.

'The findings are ground-breaking,' says Villegas. 'We found the kangaroo babies were less hyperactive, less antisocial and they even earned higher wages. This is especially significant because these were babies who were the most fragile to begin with and who came from a lower socioeconomic background. We also discovered if the father helped carry the infant, 20 years later there is a stronger family bond and less separation. Results show if you take a mother, no matter what her economic background, and give her the tools and education she needs to look after her own child, it will have the same outcomes as if she were from a higher economic status. It's a way of shortening the gap between social and educational status. This is why we say with kangaroo care, we fight inequality. We don't just save lives, we change lives.'

This story was first published on 7 February 2017 by Wellcome on mosaicscience.com

What it means to lose your sense of smell

■ Emma Young

Nick Johnson skims the lunch menu at the White Dog Cafe, a warren of little rooms and ante-rooms in Philadelphia's university district. 'Beef empanadas... I would have loved those. But all that braised beef would just get lost on me. Fish and chips I avoid: all fried foods taste the same. I'm looking at the fish tacos. I know I'll get the spicy heat and a little bit of pineapple flavour, and with the peppers and the guacamole, there'll be some mouthfeel there.'

He orders the tacos, and we get a beer that's on tap. It's called Nugget Nectar, and it's produced by the local craft brewery that Nick's worked at for the past ten years. Nugget Nectar used to be his favourite beer. 'It has a real nice balance of sweetness and hops. But now,' he says, and his face falls, 'it's a shell of its former self to me.' He can describe what it smells like: 'piney', 'citrusy', 'grapefruity'. But he can't smell it any more.

We don't think of ourselves as being particularly good smellers, especially compared with other animals. But research shows that smells can have a powerful subconscious influence on human thoughts and behaviour. People who can no longer smell – following an accident or illness – report a strong sense of loss, with impacts on their lives they could never have imagined. Perhaps we don't rank smell very highly among our senses because it's hard to appreciate what it does for us – until it's gone.

Nick, who's 34, can pinpoint the moment he lost his sense of smell. It was 9 January 2014. He was playing ice hockey with friends on the frozen pond at his parents' place in Collegeville, Pennsylvania. 'I've done it millions of times,' he says. 'I was skating backwards, slowly, and I hit a rut on the ice. My feet went out from under me. I hit the back right side of my head. I was out. I came to in the ambulance, people surrounding me, blood pouring out of my ear.' He had ruptured an eardrum and fractured his skull in three places. He had blood on his brain, and was suffering from seizures. 'I had no idea what was going on.'

After making a rapid recovery, he was cleared to drive again six weeks later and returned to work as regional sales manager at Tröegs brewery. Before long, he found himself in a meeting about a new beer. 'We were tasting it, and the others were saying, 'Can you smell the hops in the beer?'... and I couldn't. Then I tasted it. There were guys saying, 'It's got this pale biscuity flavour'... and I couldn't taste it. Then I went and tried one of the hoppier ones... and I couldn't smell it. That's when it clicked.'

The stress of the injury and all the medication perhaps explain why he didn't realise he had lost his sense of smell sooner. It came as a shock, he says. Now, though, he is acutely aware of the effects it has had.

Losing enjoyment of food and drink is a common complaint for people who lose their sense of smell. You can taste sweet, salty, bitter, sour and umami with your tongue. More complex flavours – like grapefruit or barbecued steak – depend on smell. But for Nick, as for many people who can't smell, there's another category of loss altogether.

At the time of his accident, Nick's wife was eight months pregnant with their second child. Over lunch, he says: 'I joke I can't smell my daughter's diaper. But I can't smell my daughter. She was up at 4 o'clock this morning. I was holding her, we were laying in bed. I know what my son smelt like as a little baby, as a young kid. Sometimes not so good, but he still had that great little kid smell to him. With her, I've never experienced that.'

§§§

Nick takes a deep sniff of his glass of Nugget Nectar, the beer that was once his favourite. Volatile chemicals from the liquid are drawn high up into his nostrils, to the roof of his nasal cavity, the part specialised for smelling. Then he takes a sip, and those same chemicals travel up from the back of his mouth to the same part of his nose. So far, so good.

Next, the molecules are absorbed into the mucus inside his nose. This is critical for something to be smelly: at the moment, no one can look at a molecule and say, based solely on its structure, how it will smell, or even whether it will smell at all. All we know is that for something to have a smell, its molecules must easily evaporate so they can be carried in air and inhaled, but they must also dissolve in mucus to be detected.

For a healthy person sniffing a beer, or their child, or a T-shirt belonging to their partner, exactly what happens next, leading to a perception of the beer or the person as a complex aromatic whole, is only hazily understood. Lurking within the mucus of the nose are the tips of olfactory receptor cells. These nerve cells lead directly to the brain. While we have millions of these cells, there seem to be only about 400 types, each of which binds to a specific molecule (the number is debated; some argue it could be closer to a mere 100). Based on the pattern of activation of the various receptor types, when I sniff Nugget Nectar, I recognise it as 'beer'. Nick smells nothing – the impact of his fall probably damaged or even killed his olfactory nerve cells, and his brain receives no information about the smell of his drink.

Before his injury, Nick had a highly sensitive nose. Unlike me, he would have been easily able to distinguish Nugget Nectar from other beers. That ability comes with experience. After the incoming smell signal pattern is processed, this information is sent to different parts of the brain, including regions involved in memory and emotion, as well as to the cortex, where thinking takes place. We can then quickly learn to pair patterns of receptor activation with the source of the smelly molecules.

Until recently, it was thought humans could detect perhaps only 10,000 different scents. But there's been a radical rethink, according to Joel Mainland, who's researching the fundamentals of smell at the Monell Chemical Senses Center, a world-leading institute for research on smell and taste in Philadelphia. A recent paper in the journal Science estimated that we can detect more than a trillion smells. A few problems have been raised with the methodology of that study and there's still a lot of debate about the true figure, but Mainland certainly thinks that we've underestimated our abilities.

Because of the nature of his job, Nick underwent all kinds of sensory training to improve his smell and taste. The rest of us probably have untapped potential, too. Yes, dogs are renowned for being able to track a person's scent across a field. When Mainland was a PhD student, his supervisor suggested investigating whether humans could be trained to do the same thing. It turned out they could.

Dogs have more types of olfactory receptor than we do – but as Mainland points out, cows have more than dogs (about 1,200 compared with 800) and it's not clear that cows are significantly better at smelling. The poor reputation of humans may be down to the fact that we spend comparatively little time actively sniffing and so training our sense. What difference would it make if we all put more effort into smelling the world around us?

§ § §

'Here, what does that smell of?'

George Preti's lab is lined with fume cupboards, machinery for analysing gases, and freezers packed with saliva and sweat samples and even archived T-shirts worn by people during experiments into scents produced from our underarms while we're stressed.

He waves the jar under my nose. It smells... bad. He smiles. 'Dirty laundry? It's from a Japanese group. They've isolated the smell from dirty laundry. How about this?' He removes the stopper from another jar. It smells... bad, but in a different way. 'That's 3M2H. It's one of the principal components of human underarm odour!'

The lab technician grins from a bench where he's busy preparing sample bottles for a series of new stress and fatigue experiments. 'You've just been officially Preti-labbed!' he calls.

Preti has been studying smell at Monell for decades. He specialises in the scents we humans produce. Our breath, our urine and even our blood contain molecules that smell. But the main source of body odour is the underarm region. It's rich in apocrine glands, which produce small secretions that ultimately generate underarm odour. When we're stressed, we produce more of these secretions.

Individual body odour is influenced by genes, and by one group in particular: the genes that determine which proteins combine to form your unique major histocompatibility complex (MHC). This group of proteins is on the surface of almost every cell in your body. It indicates to the immune system that a cell is 'self', and so not to be attacked. Back in 1995, researchers found that women prefer the smell of men whose suites of MHC genes are more different from their own. In 2013, a study found that people can even recognise their own scent, based on their particular MHC.

But it seems there are other smelly components of human body odour that we can detect and that can influence us. Preti and his colleagues have found that extracts of male underarm odours can not only influence female physiology, altering levels of a hormone involved in regulating the menstrual cycle, but also make women feel more relaxed and less tense. 'None of the women smelled anything different between the control and the experimental extract,' he says. 'There was an impact. But it was not conscious.'

With Pam Dalton, Preti has also investigated the impact of stress on body odour. In a study published in 2013, they collected body odour from people who'd been made to feel stressed in the lab. Another group smelled these odours while they watched videos of women doing something that might potentially be stressful – such as getting kids ready for school while trying to cook breakfast – but where the women didn't actually look stressed. ('We combed through hundreds – if not more – videos to try to find a set that worked!' Dalton remembers.)

The people watching the videos rated the women in them as being more stressed when they were exposed to the 'stressed' body odours than when they were exposed to a mild neutral fragrance or to body odour samples collected from volunteers who'd been exercising. Male viewers (but not female ones) also rated the women as being less trustworthy, less competent and less confident with the stressed odours. Yet the viewers didn't rate any of the three odours as being more or less pleasant or even as being very different from the others. The team concluded that some kind of subconscious signalling was going on.

While no one yet knows which specific chemicals convey information about our emotional state in this way, Dalton says just knowing that they exist means we can allow for them. Because she doesn't feel she produces much body odour, Dalton doesn't usually wear deodorant. However, if she knows she'll be going into a stressful situation, she puts some on – she wants to protect herself from any potentially psychologically damaging scent signals her own body may produce. She thinks we could all benefit from a better understanding of how smells can affect us: 'If we're not aware we're being influenced,' she explains, 'we can't guard against it.'

§ § §

Nick Johnson is painfully aware that he can no longer smell himself, his daughter or anyone else. It's not just because he knows what he is missing. Some people born without being able to smell may have trouble identifying the emotional states of others, says Joel Mainland. They're aware that while they rely heavily on facial expressions, for example, friends who can smell somehow seem to be picking up on signals they're missing, signals that are so powerful they can override the emotional information contained in a smile or a frown.

They'll talk about meeting up with a group of friends, Mainland explains, and one friend might say of another: 'Oh, she wasn't happy at all.' And they'll say, 'She looked happy.' And the other will say, 'Yeah, she looked happy, but she clearly was not happy.'

Estimates of just how many people can't smell usually range around a few per cent of adults. That means millions of people living without smell – some born without it, others who have lost it. Chronic nasal sinus disease is one of the most common causes of loss in younger people. Another risk stems from the fact that our olfactory receptor neurons dangle down into our nostrils, leaving them exposed to damage from environmental toxins and infections.

In older, but not elderly, people, viral infections are often to blame. Even the common cold can do it, though why it should wipe out smell in some people but not others, no one knows. By the time we get to our 70s and 80s, very few of us will have escaped significant deterioration in our sense of smell. The system has a capacity to regenerate: the nerve cells are dying all the time and being replaced. But as we get older, this process slows, and the patches of nasal tissue without any olfactory receptors get bigger.

In Nick's case, the cause was probably catastrophic damage to his olfactory receptor neurons. Heading from the nose to the brain, these neurons pass through a bony sieve-like structure. When he hit his head on the ice, the sudden movement of his brain inside his skull could have crushed or even cut them against the bone, preventing signals from his nose from reaching his brain.

Once he realised his sense of smell was gone, Nick went back to his neurologist and was surprised to be told there was nothing they could do to help. 'He told me, 'You may get your smell back in six to eight months, a year. Or you could never get it back.'

'With everything that had happened to me, I wanted an answer. And he said basically there isn't an answer.'

There are effective treatments for some people who have lost their sense of smell. If the loss is due to chronic sinus disease, you can treat that condition and reverse the smell loss – sometimes very rapidly. But for patients like Nick, there's very little that can be done. He came to Monell to ask the researchers here if they had any advice, and the main recommendation was to actively smell a few different things a couple of times a day, because there's evidence this can help to stimulate the system and may aid recovery.

Things may be different in the future. There is a team at Monell experimenting with nasal stem cells. Right now, they're investigating the most effective ways of converting these stem cells into nerve cells. The hope is that this approach will provide new olfactory receptor neurons for people whose own have been permanently damaged or defective since birth. The team hope to start animal trials in around September 2015, and if those studies go well, to move to people in five to ten years.

For now, though, Nick has to try to live with the knowledge that his sense of smell may never return.

§ § §

Life is certainly different, he says. His co-workers are supportive. But he has to rely on their descriptions of how the new beers smell and taste. He really misses the scent of the brewery, and even of the ice rink, and, now he notices what he's lost, the smell of other familiar places.

'I walk into my parents' house or my wife's family's house – and it doesn't have that smell. And I miss the ambience and the smells when there's an Eagles game, and everyone sets up grills in all the parking lots in south Philly, and grill up all kinds of crazy food items, and drink beer, hours before the game starts. Stuff you are used to... it's just gone.'

Nick used to smoke all kinds of meat and regularly barbecue for family and friends. 'I used to cook a lot more,' he says. 'This has slowed things down.' But while he can't detect complex flavours, he can still get sweet, salty, bitter, sour and umami, and also the heat of chilli. 'I love salt now. I put a lot of heat into things too, because I can get that. I put lots of cayenne in... At times, my wife will say, 'This is ridiculous, I can't even eat it!''

There are some more serious consequences of not being able to smell, though. The safety valve on his gas cooker broke recently. He was in the kitchen at a time when it must have stunk of gas, but he went to bed oblivious to the danger to his sleeping family.

The glimmer of good news for Nick is that there are some promising signs. Strong-smelling things sometimes do produce a smell sensation, though it's always the same. It used to be an awful burning-oil-type smell, he says. A few months ago, it shifted to something sweeter. This may be a sign that some kind of repair to the system is going on.

Nick says he's determined to be positive, and to live life as close as possible to the way it was before the injury. He's gone back to playing ice hockey (though he says, smiling, that he now has the best helmet money can buy). The day of our lunch at the White Dog Cafe, it's his son's third birthday. On the way home, he's going to pick up some kid-sized ice skates as a gift.

He recognises how serious the accident was, but also just how much worse things could have turned out. 'I had blood on my brain. I could have died. My outlook is: I'm glad I'm not dead. If the loss of my sense of smell is what's happened because of this, I'll take it.'

This story was first published on 4 August 2015 by Wellcome on mosaicscience.com

The doctor aiming to banish severe eye pain

■ Bryn Nelson

Razor blades. Jabbing needles. Barbed wire.

Screaming, howling, red-hot-poker-in-the-eye pain. The impulse to gouge your own eyes out or overdose on sleeping pills – anything to make the pain go away.

Blinking can be so excruciating that some people have had their eyelids partially sewn shut. One patient said the pain felt like shards of glass were jutting from her eyes. 'Imagine a knife in your eye. Forever,' wrote another.

Most causes of eye pain – a stray eyelash, a chemical burn, a dirty contact lens – are obvious and short-lived. But what happens if the source isn't immediately apparent and the agony doesn't stop? Ophthalmologists have made surprisingly little headway understanding the origins of severe and lasting eye pain. Many doctors, in fact, are outright dismissive of intense eye discomfort, deeming it of secondary importance to vision.

Patients are often written off as being hyperbolic, narcissistic or even psychiatric.

At 82, Boston ophthalmologist Perry Rosenthal hears regularly from people who are desperate for answers. Although he no longer sees patients himself, he has become the nerve centre of a small but growing network of researchers and clinicians who are defying conventional wisdom and seeking out new explanations for the often jarring disconnect between brutal symptoms and a lack of clear signs.

Rosenthal's nearly singular focus on resolving the mystery has frayed professional relationships, forced him to leave his own charity and nearly shut him out of a field in which he was once hailed as a hero. A recent burst of research, however, is prompting a new question about his unorthodox ideas: what if he's right?

§ § §

Neil Brooks' problems began at birth. Colorado-based Brooks, 51, was born long-sighted with crossed eyes and pronounced astigmatism, or irregularly shaped corneas. But it took decades for an ophthalmologist to diagnose a painful eye spasm in the ciliary muscle that helps the eye shift its focus. When his condition forced him to quit his job in commercial real estate, Brooks began searching online for a potential cause and came across a condition called accommodative spasm.

An ophthalmologist finally agreed to test Brooks' hypothesis by treating him with drops used to dilate a patient's eyes before an examination. 'About four months later, I woke up with tremendously less eye pain and significantly better vision,' Brooks says. The dilating drops had finally broken his muscle cramp.

He went back to work, this time at an online flower shop, but the long hours of computer work required him to keep increasing the frequency and strength of the eye drops. Then a new kind of eye pain appeared – a burning sensation that steadily intensified, until it felt like his corneas had been seared by the sun. Bright light became nearly intolerable. Despite trying a range of standard therapies, the new symptoms eventually eclipsed his old ones and he was again declared disabled.

Brooks talked to doctors around the world. Finally, one realised that his eye drops contained a toxic preservative known as benzalkonium chloride. The preservative, which is widely used in artificial tears and eye drops for glaucoma, has been linked to dry eye symptoms, inflammation and cell damage in multiple studies; only some manufacturers have replaced it with other chemicals or preservative-free solutions, however.

An ophthalmologist in Paris who specialised in preservative-induced toxicity confirmed the worst: six years of using the drops had ravaged Brooks' corneas at the cellular level. Brooks searched online for what to do next and eventually came upon Perry Rosenthal and his charity, Boston Foundation for Sight.

§ § §

Our clear, disc-shaped corneas serve two roles: first, as vision-enabling windows that refract incoming light and, second, as the foundations for protective films that fend off disease.

The protective 'tear film' contains three parts. First, the same kinds of cells that secrete the mucus lining our intestinal and respiratory tracts similarly coat the cornea with a thin layer of sticky mucus. Next, the lacrimal gland above each eye supplies a thicker gloss of tears. Finally, several dozen meibomian glands in the upper and lower eyelids pump out a mirror-smooth oil slick that keeps the tears from evaporating. In response to perceived threats, our eyes water or we blink, which replenishes the tears and oil. Any disruption of the tear film can result in blurred vision; repeated breaches can permanently damage the cornea.

In 2013, the nonprofit Tear Film & Ocular Surface Society launched an international public awareness campaign, Think Blink, about the importance of blinking regularly to maintain healthy eyes. The official campaign song 'Blink Around the World', a high-energy dance anthem sung by Italian pop star Sabrina, wouldn't have been out of place in the Eurovision Song Contest.

There's considerably less harmony among researchers struggling to reach a consensus on the main contributors to the condition targeted by the blinking campaign: an unwieldy catch-all of dysfunction known as 'dry eye disease'. Depending on how it's defined, studies suggest that up to one-third of the population of Japan and Taiwan may be afflicted; risk factors range from old age, smoking and low humidity to LASIK surgery and use of contact lenses or video display terminals. Women make up roughly two-thirds of all patients, for unclear reasons.

Many ophthalmologists have blamed the symptom that all patients have in common, dry-feeling eyes, on insufficient tear production or excessive tear evaporation. Other researchers suspect inflammation, and some are convinced that things still aren't that simple. The Tear Film & Ocular Surface Society has played the part of independent arbiter and recently launched its second round of workshops aiming to forge a better consensus on how dry eye disease should be defined and diagnosed. The intensive workshops are also delving deeply into subcategories such as surgery or drug-induced dry eye and abnormal pain and sensation.

Why all the bother? Studies suggest that moderate to severe dry eye can disrupt someone's quality of life just as much as hospital dialysis or moderate to severe angina, while mild dry eye can be as disruptive as severe migraines. And some researchers believe that the true spectrum of associated pain can range from occasionally scratchy eyes that could be alleviated with a simple change in lifestyle or environment to severe and unrelenting stabbing sensations that will resist nearly every intervention.

Perry Rosenthal has known two patients who committed suicide after finding no relief from the pain, and he has heard from many more who contemplated it. One woman told him that when she goes to sleep every night, she prays that she'll never wake up. 'And nothing helped. Nothing helped,' he says. 'The worst was being abandoned by the medical profession, who say it's in your mind.'

§ § §

Rosenthal grew up in northern Ontario, Canada, and considered a career as a classical pianist – his mother's preference – before choosing medicine instead. In 1960, as a 26-year-old medical resident, he founded the contact lens service at the Massachusetts Eye and Ear Infirmary in Boston. The 'father of the gas-permeable contact lens', as he is sometimes known, pioneered the introduction of lenses that didn't smother the eyes – unlike other body parts, the surface of the eye depends on air instead of blood for its oxygen supply. From research begun in a garage with two chemists, he created the Boston Lens system that was eventually bought by eye care giant Bausch + Lomb.

In the mid-1980s, as an assistant clinical professor at Harvard Medical School, Rosenthal resurrected a peculiar type of oversized contact lens first developed in Germany and Switzerland a century earlier and initially made of glass. He modernised the lenses by fashioning them from permeable acrylic and made headlines when legally blind patients suffering from corneal diseases could suddenly see after trying on what amounted to custom-built prosthetic corneas inserted with the aid of miniature plungers.

Rosenthal founded the nonprofit Boston Foundation for Sight in 1992 to help get the expensive scleral lenses to patients who had run out of other options, regardless of their ability to pay. He was featured on The Oprah Winfrey Show in 2003 and lectured at medical centres around the USA. The dome-shaped Boston Scleral Lens, as he initially called it, acts like a reservoir for artificial tears and rests on the relatively insensitive sclera – the white of the eye – instead of the hypersensitive cornea. At the charity, some of Rosenthal's patients had such severe eye damage that the lenses couldn't restore their sight. But with the lenses on, they were no longer in agony.

For Rosenthal, the turning point came in 2007, when he examined a patient from Oregon who suffered from intense eye pain and light sensitivity after she was accidentally exposed to ultraviolet radiation at work. Insufficient or quickly evaporating tears seemed like a woefully inadequate explanation for her severe symptoms. When she removed her wide-brimmed hat and dark sunglasses in the dim light of his examining room, Rosenthal watched as abundant tears spilled down her cheeks.

He fitted her with his oversized scleral lenses, which submerged the surface of each cornea in a pool of oxygen-rich artificial tears. Other specialists had accused her of lying – of malingering – to receive workers' compensation, but she responded immediately to the scleral lenses. 'It was like magic. All of a sudden, she could open her eyes wide without any light sensitivity,' Rosenthal says. Over time, however, the pain returned. 'Her response was dramatic. Why? And why did the lenses soon become much less effective? I became obsessed with the need to find an answer.'

Ophthalmologists are still unable to fit cases like this into the traditional definition of dry eye disease, Rosenthal contends, because the existing framework largely ignores the significant contribution of corneal nerves or wiring in the brain that may be sending faulty information. Eyes that feel dry, in other words, may not necessarily be dry.

§ § §

Our corneas have a far higher concentration of nerve endings than anywhere else in the body – one reason why corneal pain can be so intense. A recent description of corneal nerve anatomy, based in part on corneas from donated cadavers, suggested that 44 thickly branched nerve bundles are crammed into a space only slightly wider than the home button on an iPhone.

'It's not leaving a single point in the cornea in which there is no possibility of stimulating a pain fibre,' says Carlos Belmonte, founder of the Institute of Neurosciences in Alicante, Spain, and among the first to study the physiological basis of eye pain. The dozens of meibomian glands lining our eyelids are likewise among the most sensitive glands in the body.

One point of all those pain-producing nerves, according to evolutionary biologists, is self-preservation. The stronger the pain, the more unmistakable the message: reduce the threat now! The density of nerves in and around the cornea may help the brain send a clear warning if it senses a threat to the protective tear film that keeps our eyes healthy.

One type of pain, known as nociceptive pain, follows a relatively straightforward cause-and-effect chain of action. 'You poke me in my eye and it hurts,' says Anat Galor, an ophthalmologist at the Bascom Palmer Eye Institute at the University of Miami in Florida. Ditto for staring too long at a computer screen or exposing your eyes to a strong wind.

Belmonte's research has suggested that sensors at the ends of specialised corneal nerves can indirectly measure the thickness of the eye's tear film by sensing a drop in temperature at the corneal surface. As the tear film evaporates and exposes more of the cornea to air, Belmonte says, these sensors detect a colder surface temperature and trigger an alarm that's perceived as a dry, painful sensation. For most people, crying or blinking restores the tear film to its normal thickness, and the pain fades away. In mice missing a crucial part of the cold temperature sensor warning of dryness, Belmonte found that the normal rate of tear production fell by up to half.

A second type of pain, called neuropathic pain, transmits sensations linked to damaged, diseased or otherwise altered nerve fibres that fire spontaneously or react to light or other signals that wouldn't usually bother us. A poke in the eye hurts much more than it should; even a light touch can be excruciating. 'The nerves have kind of taken on a life of their own,' Galor says.

Rosenthal likens this system to a faulty fire alarm that's been set to go off at a certain temperature. Based on Belmonte's research, he believes damaged nerve sensors can become so hypersensitive that they send pain signals to the brain even with a normal surface temperature and an intact tear film. In other words, they continually raise a 'false dry-eye alarm' that results in chronic pain.

In what he calls his 'last hurrah', Rosenthal teamed up with Harvard University pain specialist David Borsook to lay out his ideas in a May 2015 review in the *British Journal of Ophthalmology*. The publication makes the case for ocular neuropathic pain – eye sensation rooted in dysfunctional corneal nerves – and for a related type of pain, dubbed oculofacial pain, that is instead initiated by faulty pathways in the brain. The latter condition, which Rosenthal refers to as an 'invisible, suicide-provoking eye pain', may explain some of the most extreme cases lumped under the dry-eye disease umbrella, including the woman from Oregon.

His explanation goes something like this: the patient's scleral lenses covered her tear film and temporarily quieted the pain-provoking dry eye alarm by preventing even a slight drop in the surface temperature due to tear evaporation. If the damage had been limited to her corneal nerves, maybe it would have been enough. But Rosenthal believes years of heightened pain and light sensitivity can rewire the brain, unplugging some connections and reinforcing others. Although his treatment may have initially disrupted her pain signals, he suspects her brain gradually reset its faulty connections and overrode the dampened alarm.

With this type of centralised pain, in fact, many patients cannot tolerate scleral lenses at all: 'The eyeball itself is tender,' he says. The unusually sharp, burning sensation, he believes, may originate from abnormal signals in the brain's pain-control centres that radiate out through the three branches of the trigeminal nerve supplying sensation to the head and face. 'Even though they feel it in their eyes, it's not coming from their eyes; it's projected to their eyes,' Rosenthal says. Or put another way: just as nearly one in four Danish patients in a 2010 study felt phantom pain after having their eyes amputated, some patients with oculofacial pain might still feel the intense cutting sensations even if they were to have their own eyes removed.

§ § §

None of his ideas are proven, Rosenthal concedes; they're still only hypotheses. And in any one patient, multiple problems on or under the eye's surface could be contributing to the discomfort. Nonetheless, he and his supporters are taking direct aim at lingering scepticism over whether nerve-mediated eye pain is even a real phenomenon. As Borsook points out, few doctors now doubt the existence of similar sensations in an amputated arm or leg.

'Yes, if you have your limb blown off, I see something missing and it's more tangible that you have pain and I've heard about phantom pain,' he says. 'But if you've just had a small incision in your body and you have incredible pain – it could be a tooth removal or the eye, and you still look normal – you know, what's wrong with you?'

Most ophthalmologists, Galor says, have been trained to identify evidence of cataracts or defects in the retina, not the less visible signs of abnormal nerve function. The idea that many patients with dry eye symptoms have chronic pain, then, is 'completely new and radical', she says. 'They look at it like, 'If I don't see it, it doesn't exist'.'

Patient advocates hail Rosenthal's 2009 study, 'Corneal pain without stain: is it real?' as a breakthrough in advancing the argument that debilitating pain could occur without the clinical signs detected by using fluorescent stains on the cornea and other standard tests. Citing his pioneering work, Galor calls Rosenthal 'the father of this field'.

After his initial study on chronic eye pain, however, Rosenthal says the ophthalmology community largely censored his views. Even his own charity, the Boston Foundation for Sight, was roiled by internal conflicts. When the foundation fired his son, Bill, in 2011, Rosenthal was briefly arrested for trespassing in Bill's office to gather up some belongings, according to a police report of the incident. Bill sued over his dismissal and recently reached an undisclosed settlement with the foundation. Then in 2012, after a 20-year tenure, Rosenthal was forced out of the foundation too – an abrupt firing that he alleges was linked to his focus on eye pain and what some at the charity referred to as his 'off the wall' treatments.

'Boston Foundation for Sight has had numerous disputes with Dr Rosenthal over the past many years, some of which involved his son,' responds foundation spokesperson Karen Schwartzman. 'All disputes were settled to the satisfaction of all parties in May 2015.

'With respect to Dr Rosenthal's ideas about the neuropathic origins of severe and lasting eye pain, we hope his work will encourage research on this paradigm to the benefit of patients suffering from severe eye pain.'

Based on his clinical observations at the foundation, Rosenthal wrote a paper describing 21 patients who underwent LASIK or similar laser-based surgeries and subsequently had severe eye pain lasting more than two years. After two ophthalmology journals rejected the article, he published it himself on the website of the Boston EyePain Foundation, another nonprofit that he launched in 2013 to continue his work. Since then, he has regularly posted patients' stories and railed against what he alleges is the medical community's willful suppression of mounting evidence that much of what is considered dry eye disease is instead a broken alarm mediated by faulty nerves and circuits in the brain.

David Sullivan, a Harvard ophthalmologist and founder of the Tear Film & Ocular Surface Society, is cautious about the details of Rosenthal's hypotheses, which he says are based mainly on clinical observations and may or may not be supported by further studies. 'But I think the bottom line is this whole area of pain and sensation is very, very important,' he says. And so far, he concedes, doctors don't have answers for many patients.

In his crusade for a solution, Rosenthal is finding himself increasingly accompanied by other researchers who say the general outlines of his ideas fit many of their own observations. Twenty years ago, for example, a group led by Japanese ophthalmologist Kazuo Tsubota described an unusual form of dry eye disease in which patients had no injury to the corneal surface and no drop in tear production but an unstable tear film and intense pain. 'This is a very interesting group because the symptoms, very bad. But signs? Almost nothing,' says Tsubota, who practices at Tokyo's Keio University School of Medicine. 'I love Perry Rosenthal's idea because his hypothesis can explain our findings.'

Donald Korb, an expert on meibomian gland dysfunction, says Rosenthal likewise opened his eyes to the concept of neuropathic pain. 'When I think back about how ignorant I was seven years ago, I'm appalled,' says Korb, a clinical professor of optometry at the University of California at Berkeley and a long-time friend of Rosenthal's.

'What I like about the argument that Perry has put forward is that it's shaking the field to ask the question, 'Is this a true neuropathic syndrome with pain?'' Borsook says. 'And if it is, then the current treatment of [eye] drops is not that useful.'

Eye drops are by far the most common over-the-counter remedy for dry eye, representing a multi-billion dollar global industry. The only prescription drug to win widespread regulatory approval so far, however, is ciclosporin – sold as Restasis by Dublin, Ireland-based Allergan and billed as an anti-inflammatory medication that can increase tear production. Within the past decade, more than a dozen other companies have failed in their bids to win FDA approval for a dry eye drug, although a handful of candidates are showing promise in clinical trials.

The links between corneal nerves and the brain may be more difficult to study, but Galor says most treatment strategies based on the prevailing understanding of dry eye disease haven't shown a sufficient connection between treating the signs and resolving the symptoms. Other efforts have recruited patients who may be suffering from vastly different conditions. 'We need to rethink the biology of dry eye when we design our studies,' she says.

Several groups already are. Galor says she is finding 'fantastic' success with patient-derived autologous tears, or drops made from a patient's own blood serum and filled with growth factors that might aid nerve regeneration. Others are refining the use of anticonvulsant drugs to reduce the spontaneous activity of neurons.

Sullivan is investigating a lubricating, anti-friction protein called lubricin that may reduce symptoms by preventing the tear film from becoming abnormally concentrated and unstable. Tsubota and Korb have reported promising results with goggle-like moisture chambers for some of their patients, and Korb has developed a device called LipiFlow that uses heat and pressure to relieve painfully clogged meibomian glands.

If he can secure funding, Borsook hopes to conduct an MRI imaging study of LASIK patients that might show differences in the brains of those with chronic eye pain. 'There's certainly a neuroscience interest in it,' he says, 'but the biggest thing is, can you help patients get to a point where doctors believe them?'

§ § §

That point cannot come soon enough for patients who tell remarkably similar stories about being accused of lying, of having psychiatric issues, of wasting their doctors' time.

For these patients, 'there is nothing that is more damaging to their psyche than being dismissed and invalidated by eye doctors, and they've all been through it,', Neil Brooks says. Doctor after doctor told him that because he had tears and an intact tear film, his severe pain couldn't be as bad as he described. 'And I had to come up with comebacks,' he says. 'I found myself never making progress on why my eyes might hurt because I was spending all my time trying to convince a doctor that [they did].'

In the absence of help from the medical community, Facebook has become a vital hub for many patients to share tips, encouragement and information. So have websites like the Dry Eye Zone, a collection of community forums, patient stories, blogs and an online shop run by Rebecca Petris, a former financier for the airline industry turned off-the-grid homesteader in rural Washington State.

Since her own LASIK surgery in 2001, Petris has faced more than a decade of experimental therapies and corrective surgeries to resolve chronic vision and dry eye symptoms, although she counts herself lucky that she doesn't have severe stabbing sensations.

With no common language for chronic pain, she says, patients who aren't taken seriously may use increasingly dramatic descriptions, only to have the strategy backfire when doctors view them with mounting suspicion. Petris now advises patients to go to their appointments armed with standardised questionnaires like the Ocular Surface Disease Index, a 12-question survey on the severity of symptoms that's now available as a smartphone app. 'The difficulty of putting pain in terms that doctors can understand is huge,' she says. 'I don't know how you get over that.'

For relief, Petris sometimes recommends one of the Dry Eye Shop remedies she's curated over the years. Sometimes she becomes an impromptu therapist instead, urging callers to hang on and seek help before they take their own lives. 'People find me on the internet because no one's listening to them,' she says. 'They're at the end of their rope.'

Both patients and researchers say the message may finally be getting through. Every big advancement in medicine seems to follow the same cycle, Neil Brooks says: people don't believe it and then an avant-garde researcher questions the prevailing truth and pushes the envelope to find out what's really happening. 'Perry Rosenthal, to my knowledge, was at the vanguard of saying, 'What if these people aren't crazy?'' he says.

After reaching out to Rosenthal, Brooks agreed to see whether a pair of custom-built scleral lenses might relieve his symptoms. He endured long and painful fitting sessions but immediately noticed a difference when he left with a new pair. 'There are no miracles when you've gotten to the point where my eyes are, but it was the single best, most effective treatment,' Brooks says. The lenses not only reduced his pain but also made them less sensitive to the dry, bright and windy conditions back home in Colorado.

The relief wouldn't last. Because Brooks had undergone multiple surgeries to correct his crossed eyes, three specialists told him that the suction of the scleral lenses was putting his eyes at risk for even more damage. A 'life-changing' transformation was once again cut short. 'So I went from [being] that guy who could do things back to being the guy who can't,' he says. At the same time, the 'invisible disease' was wreaking havoc with his personal life: a simple disagreement over barking dogs that continually interrupted his desperately needed sleep – he didn't seem sick to his neighbours – spiralled out of control and eventually forced him out of his house.

Brooks made one last big push to reclaim his life: he sold his house, put everything in storage and immersed himself in the ultra-humid tropics of Central America. For the first ten weeks, life was great; the high humidity helped to keep his tear film intact. Then the effect began to wear off, and the pain returned. The body is a self-regulating mechanism, Rosenthal told him. Eventually, his brain may have instructed his lacrimal glands to acclimate to the new environment by reducing their tear production.

He has since retreated to Colorado, where he lives with relatives. Nearly two years later, his remaining strategy is avoidance. 'I don't do anything,' Brooks says. 'If I am lucky, I get the dog out for a walk four or five times a week.'

Perry Rosenthal is still in battle mode and pessimistic about whether ophthalmologists are truly motivated to get at the root cause of chronic eye pain. Even his supporters aren't in full agreement on the chain of events triggering the sensitivity. But more researchers are at least coalescing around the idea that neuropathic eye pain may be caused by an accident, disease, drug use or surgery.

What Brooks has left, he says himself, is hope. 'A once full and big and wonderful life has been reduced to next to nothing,' he says. 'And I know that's true of a lot of other people, and I have to tell myself the same thing I would tell them, which is don't give up – the same doctor you saw today might read a paper tomorrow that lets him help you next week.'

This story was first published on 8 September 2015 by Wellcome on mosaicscience.com

Allergies — a defence against noxious chemicals?

■ Carl Zimmer

For me, it was hornets.

One summer afternoon when I was 12, I ran into an overgrown field near a friend's house and kicked a hornet nest the size of a football. An angry squadron of insects clamped onto my leg; their stings felt like scorching needles. I swatted the hornets away and ran for help, but within minutes I realised something else was happening. A constellation of pink stars had appeared around the stings. The hives swelled, and new ones began appearing farther up my legs. I was having an allergic reaction.

My friend's mother gave me antihistamines and loaded me into her van. We set out for the county hospital, my dread growing as we drove. I was vaguely aware of the horrible things that can happen when allergies run amok. I imagined the hives reaching my throat and sealing it shut.

I lived to tell the tale: my hives subsided at the hospital, leaving behind a lingering fear of hornets. But an allergy test confirmed that I was sensitive to the insects. Not to honey bees or wasps or yellow jackets. Just the particular type of hornet that had stung me. The emergency room doctor said I might not be so fortunate the next time I encountered a nest of them. She handed me an EpiPen and told me to ram the syringe into my thigh if I was stung again. The epinephrine would raise my blood pressure, open my airway – and perhaps save my life. I've been lucky: that afternoon was 35 years ago, and I haven't encountered a hornet's nest since. I lost track of that EpiPen years ago.

Anyone with an allergy has their origin story, a tale of how they discovered that their immune system goes haywire when some arbitrarily particular molecule gets into their body. There are hundreds of millions of these stories. In the USA alone, an estimated 18 million people suffer from hay fever, and food allergies affect millions of American children. The prevalence of allergies in many other countries is rising. The list of allergens includes – but is not limited to – latex, gold, pollen (ragweed, cockleweed and pigweed are especially bad), penicillin, insect venom, peanuts, papayas, jellyfish stings, perfume, eggs, the faeces of house mites, pecans, salmon, beef and nickel.

Once these substances trigger an allergy, the symptoms can run the gamut from annoying to deadly. Hives appear, lips swell. Hay fever brings sniffles and stinging eyes; allergies to food can cause vomiting and diarrhoea. For an unlucky minority, allergies can trigger a potentially fatal whole-body reaction known as anaphylactic shock.

The collective burden of these woes is tremendous, yet the treatment options are limited. EpiPens save lives, but the available long-term treatments offer mixed results to those exhausted by an allergy to mould or the annual release of pollen. Antihistamines can often reduce sufferers' symptoms, but these drugs also cause drowsiness, as do some other treatments.

We might have more effective treatments if scientists understood allergies, but a maddening web of causes underlies allergic reactions. Cells are aroused, chemicals released, signals relayed. Scientists have only partially mapped the process. And there's an even bigger mystery underlying this biochemical web: why do we even get allergies at all?

'That is exactly the problem I love,' Ruslan Medzhitov told me recently. 'It's very big, it's very fundamental, and completely unknown.'

Medzhitov and I were wandering through his laboratory, which is located on the top floor of the Anlyan Center for Medical Research and Education at the Yale School of Medicine. His team of postdocs and graduate students were wedged tight among man-sized tanks of oxygen and incubators full of immune cells. 'It's a mess, but a productive mess,' he said with a shrug. Medzhitov has a boxer's face – massive, circular, with a broad, flat nose – but he spoke with a soft elegance.

Medzhitov's mess has been exceptionally productive. Over the past 20 years, he has made fundamental discoveries about the immune system, for which he has been awarded a string of major prizes. Last year he was the first recipient of the €4 million Else Kröner Fresenius Award. And though Medzhitov hasn't won a Nobel, many of his peers think he should have: in 2011, 26 leading immunologists wrote to Nature protesting that Medzhitov's research had been overlooked for the prize.

Now Medzhitov is turning his attention to a question that could change immunology yet again: why do we get allergies? No one has a firm answer, but what is arguably the leading theory suggests that allergies are a misfiring of a defence against parasitic worms. In the industrialised world, where such infections are rare, this system reacts in an exaggerated fashion to harmless targets, making us miserable in the process.

Medzhitov thinks that's wrong. Allergies are not simply a biological blunder. Instead, they're an essential defence against noxious chemicals – a defence that has served our ancestors for tens of millions of years and continues to do so today. It's a controversial theory, Medzhitov acknowledges. But he's also confident that history will prove him right. 'I think the field will go around in that stage where there's a lot of resistance to the idea,' he told me. 'Until everybody says, 'Oh yeah, it's obvious. Of course it works that way.''

§ § §

The physicians of the ancient world knew about allergies. Three thousand years ago, Chinese doctors described a 'plant fever' that caused runny noses in autumn. There is evidence that the Egyptian pharaoh Menes died from the sting of a wasp in 2641 BCE. Two and a half millennia later, the Roman philosopher Lucretius wrote, 'What is food to one is to others bitter poison.'

But it was a little more than a century ago when scientists realised that these diverse symptoms are different heads on the same hydra. By then researchers had discovered that many diseases are caused by bacteria and other pathogens, and that we fight these invaders with an immune system – an army of cells that can unleash deadly chemicals and precisely targeted antibodies. They soon realised that the immune system can also cause harm. In the early 1900s, the French scientists Charles Richet and Paul Portier were studying how toxins affect the body. They injected small doses of poison from sea anemones into dogs, then waited a week or so before delivering an even smaller dose. Within minutes, the dogs went into shock and died. Instead of protecting the animals from harm, the immune system appeared to make them more susceptible.

Other researchers observed that some medical drugs caused hives and other symptoms. And this sensitivity increased with exposure – the opposite of the protection that antibodies provided against infectious diseases. The Austrian doctor Clemens von Pirquet wondered how it was that substances entering the body could change the way the body reacted. To describe this response, he coined the word 'allergy', from the Greek words allos ('other') and ergon ('work').

In the decades that followed, scientists discovered that the molecular stages of these reactions were remarkably similar. The process begins when an allergen lands on one of the body's surfaces – skin, eye, nasal passage, mouth, airway or gut. These surfaces are loaded with immune cells that act as border sentries. When a sentry encounters an allergen, it first engulfs and demolishes the invader, then decorates its outer surface with fragments of the substance. Next the cell locates some lymph tissue. There it passes on the fragments to other immune cells, which produce a distinctive fork-shaped antibody, known as immunoglobulin E, or IgE.

These antibodies will trigger a response if they encounter the allergen again. The reaction begins when an antibody activates a component of the immune system known as a mast cell, which then blasts out a barrage of chemicals. Some of these chemicals latch onto nerves, triggering itchiness and coughing. Sometimes mucus is produced. Airway muscles can contract, making it hard to breathe.

This picture, built up in labs over the past century, answered the 'how?' part of the allergies mystery. Left unanswered, however, was 'why?' And that's surprising, because the question had a pretty clear answer for most parts of the immune system. Our ancestors faced a constant assault of pathogens. Natural selection favoured mutations that helped them fend off these attacks, and those mutations accumulated to produce the sophisticated defences we have today.

It was harder to see how natural selection could have produced allergies. Reacting to harmless things with a huge immune response probably wouldn't have aided the survival of our ancestors. Allergies are also strangely selective. Only some people have allergies, and only some substances are allergens. Sometimes people develop allergies relatively late in life; sometimes childhood allergies disappear. And for decades, nobody could even figure out what IgE was for. It showed no ability to stop any virus or bacteria. It was as if we evolved one special kind of antibody just to make us miserable.

One early clue came in 1964. A parasitologist named Bridget Ogilvie was investigating how the immune system repelled parasitic worms, and she noticed that rats infected with worms produced large amounts of what would later be called IgE. Subsequent studies revealed that the antibodies signalled the immune system to unleash a damaging assault on the worms.

Parasitic worms represent a serious threat – not just to rats, but to humans too. Hookworms can drain off blood from the gut. Liver flukes can damage liver tissue and cause cancer.

Tapeworms can cause cysts in the brain. More than 20 per cent of all people on Earth carry such an infection, most of them in low-income countries. Before modern public health and food safety systems, our ancestors faced a lifelong struggle against these worms, as well as ticks and other parasitic animals.

During the 1980s, several scientists argued forcefully for a link between these parasites and allergies. Perhaps our ancestors evolved an ability to recognise the proteins on the surface of worms and to respond with IgE antibodies. The antibodies primed immune system cells in the skin and gut to quickly repel any parasite trying to push its way in. 'You've got about an hour to react very dramatically in order to reduce the chance of these parasites surviving,' said David Dunne, a parasitologist at the University of Cambridge.

According to the worm theory, the proteins of parasitic worms are similar in shape to other molecules we regularly encounter in our lives. If we encounter those molecules, we mount a pointless defence. 'Allergy is just an unfortunate side-effect of defence against parasitic worms,' says Dunne.

§ § §

When he was an immunologist in training, Medzhitov was taught the worm theory of allergies. But ten years ago he started to develop doubts. 'I was seeing that it doesn't make sense,' he said. So Medzhitov began thinking about a theory of his own.

Thinking is a big part of Medzhitov's science. It's a legacy of his training in the Soviet Union in the 1980s and 1990s, when universities had little equipment and even less interest in producing good scientists. For his undergraduate degree,

Medzhitov went to Tashkent State University in Uzbekistan. Every autumn the professors sent the students out into the cotton fields to help take in the harvest. They worked daily from dawn to dusk. 'It was terrible,' said Medzhitov. 'If you don't do that, you get expelled from college.' He recalls sneaking biochemistry textbooks into the fields – and being reprimanded by a department chair for doing so.

Graduate school wasn't much better. Medzhitov arrived at Moscow State University just as the Soviet regime collapsed. The university was broke, and Medzhitov didn't have the equipment he needed to run experiments. 'I was basically spending all of my time reading and thinking,' Medzhitov told me.

Mostly, he thought about how our bodies perceive the outside world. We can recognise patterns of photons with our eyes and patterns of air vibrations with our ears. To Medzhitov, the immune system was another pattern recognition system – one that detected molecular signatures instead of light or sound.

As Medzhitov searched for papers on this subject, he came across references to a 1989 essay written by Charles Janeway, an immunologist at Yale, titled 'Approaching the Asymptote? Evolution and revolution in immunology'. Medzhitov was intrigued and used several months' of his stipend to buy a reprint of the paper. It was worth the wait, because the paper exposed him to Janeway's theories, and those theories would change his life.

At the time, Janeway was arguing that antibodies have a big drawback: it takes days for the immune system to develop an effective antibody against a new invader. He speculated that the immune system might have another line of defence that could offer faster protection. Perhaps the immune system could use a pattern-recognition system to detect bacteria and viruses quickly, allowing it to immediately launch a response.

Medzhitov had been thinking about the same thing, and he immediately emailed Janeway. Janeway responded, and they began an exchange that would ultimately bring Medzhitov to New Haven, Connecticut, in 1994, to become a postdoctoral researcher in Janeway's lab. (Janeway died in 2003.)

'He turned out to speak very little English, and had almost no experience in a wet laboratory,' says Derek Sant'Angelo, who worked in the lab at the time. Sant'Angelo, now at the Robert Wood Johnson Medical School in New Jersey, recalls coming across Medzhitov at the bench one night. In one hand, Medzhitov held a mechanical pipette. In the other hand, he held a tube of bacteria. Medzhitov needed to use the pipette to remove a few drops of bacteria from the tube and place them on a plate on the lab bench in front of him. 'He was slowly looking back and forth from the pipette down to the plate to the bacteria,' says Sant'Angelo. 'He knew in theory that the pipette was used to put the bacteria on the plate. But he simply had absolutely no idea how to do it.'

Medzhitov still marvels that Janeway agreed to work with him. 'I think that the only reason that he took me in his lab is that nobody else wanted to touch this idea,' he recalled.

With help from Sant'Angelo and other members of the lab, Medzhitov learned very quickly. Soon he and Janeway discovered a new class of sensor on the surface of a certain kind of immune cell. Confronted with an invader, the sensors would clasp onto the intruder and trigger a chemical alarm that promoted other immune cells to search the area for pathogens to kill. It was a fast, accurate way to sense and remove bacterial invaders.

Medzhitov and Janeway's discovery of the sensors, now known as toll-like receptors, revealed a new dimension to our immune defences, and has been hailed as a fundamental principle of immunology. It also helped solve a medical mystery.

Infections sometimes produce a catastrophic body-wide inflammation known as sepsis. It is thought to strike around a million people a year in the USA alone, up to half of whom die. For years, scientists thought that a bacterial toxin might cause the immune system to malfunction in this way – but sepsis is actually just an exaggeration of one of the usual immune defences against bacteria and other invaders. Instead of acting locally, the immune system accidentally responds throughout the body. 'What happens in septic shock is that these mechanisms become activated much more strongly than necessary,' said Medzhitov. 'And that's what kills.'

Medzhitov isn't driven to do science to cure people; he's more interested in basic questions about the immune system. But he argues that cures won't be found if researchers have the wrong answers for basic questions. Only now that scientists have a clear understanding of the biology underlying sepsis can they develop treatments that target the real cause of the condition – the over-reaction of the toll-like receptors. (Tests are ongoing, and the results so far are promising). 'Thirty years ago, it was, 'Whatever causes septic shock is bad.' Well, now we know it's not,' said Medzhitov.

§ § §

Medzhitov kept thinking after he and Janeway discovered toll-like receptors. If the immune system has special sensors for bacteria and other invaders, perhaps it had undiscovered sensors for other enemies. That's when he started thinking about parasitic worms, IgE and allergies. And when he thought about them, things didn't add up.

It's true that the immune system makes IgE when it detects parasitic worms. But some studies suggest that IgE isn't actually essential to fight these invaders. Scientists have engineered mice that can't make IgE, for instance, and have found that the animals can still mount a defence against parasitic worms. And Medzhitov was sceptical of the idea that allergens mimic parasite proteins. A lot of allergens, such as nickel or penicillin, have no possible counterpart in the molecular biology of a parasite.

The more Medzhitov thought about allergens, the less important their structure seemed. Maybe what ties allergens together was not their shape, but what they do.

We know that allergens often cause physical damage. They rip open cells, irritate membranes, slice proteins into tatters. Maybe, Medzhitov thought, allergens do so much damage that we need a defence against them. 'If you think of all the major symptoms of allergic reactions – runny noses, tears, sneezing, coughing, itching, vomiting and diarrhoea – all of these things have one thing in common,' said Medzhitov. 'They all have to do with expulsion.' Suddenly the misery of allergies took on a new look. Allergies weren't the body going haywire; they were the body's strategy for getting rid of the allergens.

As Medzhitov explored this possibility, he found that the idea had surfaced from time to time over the years, only to be buried again. In 1991, for example, the evolutionary biologist Margie Profet argued that allergies fought toxins. Immunologists dismissed the idea, perhaps because Profet was an outsider. Medzhitov found it hugely helpful. 'It was liberating,' he said.

Together with two of his students, Noah Palm and Rachel Rosenstein, Medzhitov published his theory in Nature in 2012. Then he began testing it. First he checked for a link between damage and allergies. He and colleagues injected mice with PLA2, an allergen that's found in honey-bee venom and tears apart cell membranes. As Medzhitov had predicted, the animals' immune systems didn't respond to PLA2 itself. Only when PLA2 ripped open cells did the immune system produce IgE antibodies.

Another prediction of Medzhitov's theory was that these antibodies would protect the mice, rather than just make them ill. To test this, Medzhitov and his colleagues followed their initial injection of PLA2 with a second, much bigger dose. If the animals had not previously been exposed to PLA2, the dose sent their body temperature plunging, sometimes fatally. But the mice that had been exposed marshalled an allergic reaction that, for reasons that aren't yet clear, lessened the impact of the PLA2.

Medzhitov didn't know it, but on the other side of the country another scientist was running an experiment that would provide even stronger support for his theory. Stephen Galli, chair of the Pathology Department at Stanford University School of Medicine, had spent years studying mast cells, the enigmatic immune cells that can kill people during allergic reactions. He suspected mast cells may actually help the body. In 2006, for example, Galli and colleagues found that mast cells destroy a toxin found in viper venom. That discovery led Galli to wonder, like Medzhitov, whether allergies might be protective.

To find out, Galli and colleagues injected one to two stings' worth of honey-bee venom into mice, prompting an allergic reaction. Then they injected the same animals with a potentially lethal dose, to see if the reaction improved the animal's chance of survival. It did. What's more, when Galli's team injected the IgE antibodies into mice that had never been exposed to the venom, those animals were also protected against a potentially lethal dose.

Medzhitov was delighted to discover Galli's paper in the same issue of Immunity that carried his own. 'It was good to see that somebody got the same results using a very different model. That's always reassuring,' Medzhitov told me.

Still, the experiments left a lot unanswered. How precisely did the damage caused by the bee venom lead to an IgE response? And how did IgE protect the mice? These are the kinds of questions that Medzhitov's team is now investigating. He showed me some of the experiments when I visited again last month. We sidled past a hulking new freezer blocking a corridor to slip into a room where Jaime Cullen, a researcher associate in the lab, spends much of her time. She put a flask of pink syrup under a microscope and invited me to look. I could see a flotilla of melon-shaped objects.

'These are the cells that cause all the problems,' said Medzhitov. I was looking at mast cells, the key agents of allergic reactions. Cullen is studying how IgE antibodies latch onto mast cells and prime them to become sensitive – or, in some cases, oversensitive – to allergens.

Medzhitov predicts that these experiments will show that allergen detection is like a home-alarm system. 'You can detect a burglar, not by recognising his face, but by a broken window,' he said. The damage caused by an allergen rouses the immune system, which gathers up molecules in the vicinity and makes antibodies to them. Now the criminal has been identified and can be more easily apprehended next time he tries to break in.

Allergies make a lot more sense in terms of evolution when seen as a home-alarm system, argues Medzhitov. Toxic chemicals, whether from venomous animals or plants, have long threatened human health. Allergies would have protected our ancestors by flushing out these chemicals. And the discomfort our ancestors felt when exposed to these allergens might have led them to move to safer parts of their environment.

Like many adaptations, allergies weren't perfect. They lowered the odds of dying from toxins but didn't eliminate the risk. Sometimes the immune system overreacts dangerously, as Richet and Protier discovered when the second dose of anemone allergen killed the dogs they were experimenting on. And the immune system might sometimes round up a harmless molecular bystander when it responded to an allergy alarm. But overall, Medzhitov argues, the benefits of allergies outstripped their drawbacks.

That balance shifted with the rise of modern Western life, he adds. As we created more synthetic chemicals, we exposed ourselves to a wider range of compounds, each of which could potentially cause damage and trigger an allergic reaction. Our ancestors could avoid allergens by moving to the other side of the forest, but we can't escape so easily. 'In this particular case, the environment we'd have to avoid is living indoors,' said Medzhitov.

Scientists are taking this theory very seriously. 'Ruslan is one of the most distinguished immunologists in the world,' said Galli. 'If he thinks there's validity to this idea, I think it gets a lot of traction.'

Dunne, on the other hand, is sceptical about the idea that Medzhitov's theory explains all allergies. Medzhitov is underestimating the huge diversity of proteins that Dunne and others are finding on the surface of worms – proteins that could be mimicked by a huge range of allergens in the modern world. 'My money's more on the worm one,' he said.

§ § §

Over the next few years, Medzhitov hopes to persuade sceptics with another experiment. It's unlikely to end the debate, but positive results would bring many more people over to his way of thinking. And that might eventually lead to a revolution in the way we treat allergies.

Sitting on Cullen's lab bench is a plastic box that houses a pair of mice. There are dozens more of these boxes in the basement of their building. Some of the mice are ordinary, but others are not: using genetic engineering techniques, Medzhitov's team has removed the animals' ability to make IgE. They can't get allergies.

Medzhitov and Cullen will be observing these allergy-free mice for the next couple of years. The animals may be spared the misery of hay fever caused by the ragweed pollen that will inevitably drift into their box on currents of air. But Medzhitov predicts they will be worse off for it. Unable to fight the pollen and other allergens, they will let these toxic molecules pass into their bodies, where they will damage organs and tissues.

'It's never been done before, so we don't know what the consequences will be,' says Medzhitov. But if his theory is right, the experiment will reveal the invisible shield that allergies provide us.

Even if the experiment works out just as he predicts, Medzhitov doesn't think his ideas about allergies will win out as quickly as his ideas about toll-like receptors. The idea that allergic reactions are bad is ingrained in the minds of physicians. 'There's going to be more inertia,' he said.

But understanding the purpose of allergies could lead to dramatic changes in how they're treated. 'One implication of our view is that any attempt to completely block allergic defences would be a bad idea,' he said. Instead, allergists should be learning why a minority of people turn a protective response into a hypersensitive one. 'It's the same as with pain,' said Medzhitov. 'No pain at all is deadly; normal pain is good; too much pain is bad.'

For now, however, Medzhitov would just be happy to get people to stop seeing allergies as a disease, despite the misery they cause. 'You're sneezing to protect yourself. The fact that you don't like the sneezing, that's tough luck,' he said, with a slight shrug. 'Evolution doesn't care how you feel.'

This story was first published on 7 April 2015
by Wellcome on mosaicscience.com

Why pharma may be going slow on the male pill

■ Andy Extance

Had there been a male contraceptive pill in 1976, I probably wouldn't be here to write this. That was the year when, after my mum – may she rest in peace – had been on the pill for 12 years, health worries made her doctor tell her to come off it. 'She said to the doctor, 'I'll get pregnant',' my dad recalls. 'And within a very short while, she was.' He explains, much to my discomfort, that although my parents switched to condoms, I was conceived because 'sometimes you feel reckless'. But if a male pill had existed, my dad says, he'd definitely have used it.

So why didn't it exist? It certainly wasn't because of a lack of scientific interest. Gregory Pincus, who co-invented the female contraceptive pill, first tested the same hormonal approach on men in 1957, and various hormonal and non-hormonal methods have been explored since. And although attitudes among those who might use a male pill were once thought to be a daunting obstacle, it's now clear that many men want a new option.

Despite this, we're still waiting. Developing a method that men would accept has brought decades of frustration, yet researchers are as confident as they can be that they're close to overcoming the scientific barriers. But, crucially, drug makers' commitment to contraceptives has always been tentative, particularly when it comes to products for men – and today, the whole contraceptive industry is struggling. Now, the multimillion-dollar question seems to be: Who is actually going to make the male pill happen?

§ § §

In the 1970s, when my dad might have used a contraceptive pill, prospects seemed better in some ways. Male fertility control was an active research field, with governments backing various ideas to limit overcrowding on Earth. One product he might have been interested in – a non-hormonal drug called gossypol – was being tested on a scale that has never been matched since. At the UN's 1974 World Population Conference, Elsimar Coutinho, today a famous sex and fertility doctor in Brazil, was promoting the drug, which he was testing on men at the Federal University of Bahia. However, attitudes surrounding sex and reproduction can be unpredictable, and not everyone was convinced of its worth.

'The conference hall was full of women,' Coutinho says on the phone, his gravelly voice matching his website's picture of a suave doctor with slicked-back grey hair. 'I was going to tell them, 'Now you don't have to take pills if you don't want."' Yet, having determined their own fertility through the contraceptive pill for little more than a decade, his female audience were determined not to relinquish control. 'To my surprise, I was shouted down and booed out.'

Despite such reactions, poorer countries with fast-growing populations found gossypol appealing because it could be extracted cheaply from cotton farming waste. Coutinho had first seen its potential while visiting Brazilian farmers who fed cotton plant debris to their bulls. 'The bulls were having sex more often, the farmers thought it was good for sexual prowess,' he recalls. But actually, the bulls were not making enough sperm and were therefore still surrounded by receptive, non-pregnant cows – and just doing what came naturally.

From the 1960s onwards, Coutinho worked on contraception with the Chinese government, which in 1972 ran trials with 8,806 men taking gossypol pills. Daily doses successfully reduced the men's sperm count enough to satisfy the researchers, but side-effects were a cause for concern. One notable problem was that 66 of the men had low potassium in their blood. More importantly, sperm levels in many men didn't return to normal when they stopped taking the drug.

Researchers therefore conducted tests for years longer, showing in rats that gossypol doesn't just stop sperm moving, but also damages the lining of epididymis ducts, which store sperm made by the testicles. Eventually, an October 1986 symposium in Wuhan, China – whose sponsors included the Chinese government and the World Health Organization (WHO) – concluded that gossypol was 'of little interest'.

'You may call it a problem, but we saw it as a solution,' Coutinho tells me. He felt that the fact it could be irreversible made gossypol a potential alternative to surgical vasectomy. He joined with an international team of scientists to conduct further trials, the last of whose results were published in 2000. They found no problems with potassium, putting the effects seen in China down to poor diet.

The researchers therefore applied to the Brazilian government for permission to sell the drug, needing to overcome the strong influence of the Roman Catholic Church, which forbids artificial contraception. On 14 June 2001, Josimar Henrique da Silva, founder of the Brazilian drug company Hebron, which was hoping to commercialise gossypol, wrote to Coutinho. 'I'm working at the Ministry of Health in such a way as not to create more obstacles,' Coutinho reads from the letter. 'I can't fight against them. Give me two more weeks.'

Coutinho never heard from da Silva again. The gossypol contraceptive saga ended in failure after more than three decades. Coutinho mischievously suggests Ministry machismo may have been a contributing factor. 'We worked on this for many years and realised men are very afraid of losing virility,' he says. 'Maybe those judging our application were amongst them.'

By contrast, my dad apparently wouldn't have seen a male pill as a threat to his virility, and I too would be interested in rather than threatened by a new male contraceptive – I believe it would benefit, rather than harm, the sex my partner and I have. Are we unusual in that?

§ § §

Actually, plenty of men are interested in a male pill. In 2005, researchers in Germany published a study asking over 9,000 men from nine countries on four continents whether they'd use a contraceptive method 'capable of preventing sperm production'. Over half were willing, the proportion ranging from three-tenths to seven-tenths depending on the country.

Other surveys report similar attitudes. In 2011, Susan Walker at Anglia Ruskin University in Chelmsford, UK, published a small study including 54 men in an anonymous town in England. Twenty-six of them said yes, they would take it. 'They were not concerned about losing fertility – as long as they could be sure of regaining it,' Walker stresses.

The remainder, who split between responding no and don't know, showed some gender-based reluctance. 'It's a strange idea,' one man said. 'I'm so used to women taking the pill.' Those who were unsure were more concerned about side-effects, Walker notes. 'They said, 'I've seen what the pill does to my girlfriend', 'What would the long-term effects on my fertility be?', 'Could I be sure my fertility would return?', that kind of really quite sensible concern.'

The survey also included 134 women, roughly half of whom would let their partners use a male pill. However, more than half were worried that men would forget to take the pill regularly, whereas just one in six of the men had this worry. 'Of course, women have the experience of having to remember to take the pill,' Walker says. One study from 1996, in which 103 women were given electronic pill dispensers that monitored what they'd taken, found that they missed 2.6 pills per month on average.

'The general concept is that there are men out there that would use it,' says Richard Anderson, a professor of reproductive science at the University of Edinburgh. And some women would trust them – although often media coverage might suggest otherwise. 'Whenever there's a study published, a radio journalist will walk up and down the high street in their local town and ask women whether they'd trust a man to take a pill, and of course they all run for the hills. But if you ask a woman if they would trust their partner, who they share children, their bank account, and a bed every night with, then you're going to get a different answer.'

In 1995 and 1996, researchers including Anderson interviewed 1,829 men across four cities: Edinburgh, Cape Town, Shanghai and Hong Kong. White men in Cape Town were most eager, with four-fifths saying they would at least probably use a male hormonal contraceptive pill. Hong Kong residents were least keen, with two-fifths saying that they would definitely or probably take a pill. Fewer were interested if the drug was injected – three-fifths of white men in Cape Town and a third of men in Hong Kong, for example. 'It's not going to be right for everybody,' Anderson says. 'The whole concept is to provide a range of options so that individuals can find what suits them best.'

A photo of Anderson's reveals the injection question's importance. In it, a woman is grinning as she depresses the plunger on the hormone-filled syringe she's injecting into her husband's naked bottom. This was the method used in the first WHO-backed clinical trial that Anderson was involved in, back in 1991. 'It was proof that you could use a hormonal method to produce real contraceptive efficacy,' Anderson says. This trial also helped show that male contraceptives didn't need to cut sperm count to zero. With anything upwards of 15 million sperm per millilitre considered normal, the trial set its maximum threshold at 3 million per millilitre. The consensus today is that 'anything below 1 million per millilitre is going to provide pretty good contraception,' Anderson says.

§ § §

A single crumpled piece of A4 paper on an almost-bare wall in Anderson's office illustrates how hormonal male contraceptives work, reducing men to brain and balls. In the brain, it picks out the hypothalamus and pituitary gland. In the testicles it shows cells that make testosterone, and the tubules they neighbour, where sperm are made. Progestogen hormones like those used in female pills can stop the glands in a man's brain making luteinising hormone and follicle-stimulating hormone. The absence of these hormones stops the man's testicles producing sperm – but it also stops them producing testosterone. So testosterone replacement is given along with progestogens, to avoid undesirable effects like weaker muscles and lessened sex drive.

The Edinburgh scientists' various trials have long attracted media attention. "100% success' for male pill trial,' trumpeted the BBC in 2000, reporting on the suppression of sperm production in 30 men – reportedly without side-effects – from a combined progestogen pill and testosterone-releasing implant. Both hormones came from the Dutch drug company Organon, which, after what Anderson calls 'a lot of persuading', began to pay attention.

Eventually Organon teamed up with Germany's Schering on a larger clinical trial in 2003–04. Researchers gave 297 men progestogen implants Organon was developing for women and injections of a Schering testosterone product. They gave around 52 more men placebos – all the participants were also using other contraception – and monitored their sperm count. For almost nine-tenths of the men on the hormonal contraceptive, sperm counts fell below the million mark, and once the trial was over they all recovered normal fertility after around four months. But not everything was ideal. More men taking hormones suffered 'adverse events' like acne, sweating, and effects on weight, mood and sex drive than the placebo group. Some of these were more serious and even life-threatening, including one attempted suicide.

This would be the pinnacle of Big Pharma's interest. Between running the trial and publishing results in 2008, Schering was bought by German rival Bayer, which ended work on the subject. Organon likewise ended its interest, which Herjan Coelingh Bennink, global executive vice-president in the company's reproductive medicine programme until 2000, believes is partly because this work lacked support in the company. Encouraged by the survey done by Anderson and his colleagues, Coelingh Bennink had pushed the approach, and helped design the joint trial. But – echoing Coutinho's account of his experience with gossypol – among Organon's most senior leaders attitudes were not as open as elsewhere.

'At board level it was only middle-aged white males,' Coelingh Bennink recalls. 'I tried to explain how important it could be, but they never got further than saying to each other, 'Would you do it?' 'No, I wouldn't do it.' It was not considered male behaviour to take responsibility for contraception.'

On leaving Organon, Coelingh Bennink founded Pantarhei Bio. There, he has overseen development of an improved female contraceptive pill that offers lessons for the male version. When women start using hormonal contraceptives, there's a possibility that blood clots can form. While the risk is very low, it does happen and can lead to serious complications – for the women and for the drug companies. For example, Bayer is estimated to have paid around $2 billion to women who have sued it over such blood clots. Likewise Merck and Johnson & Johnson have paid millions of dollars to settle similar cases brought against them. The new drug 'most likely' doesn't cause clots, Coelingh Bennink says.

In 2016, the new female pill will enter large-scale phase III human trials – the ultimate test of whether drugs work – to determine whether government regulators will approve its sale. But Coelingh Bennink estimates these will cost €50–100 million, and Pantarhei doesn't have the money. Instead, it has sold rights to the drug to a Belgian company, Mithra Pharmaceuticals, who are running the necessary clinical trials.

Getting to this point has taken Pantarhei 14 years, and finding a partner prepared to risk large-scale testing has been one of the hardest parts. Contraceptive drug companies have all drastically cut funding for new products, Coelingh Bennink says. 'It's a disastrous world to develop drugs in. It's much more profitable to develop another cancer drug. Contraceptives are a retail business – it's a matter of selling a lot, and profit is low.'

US-based Transparency Market Research estimates that people across the world spent almost $16 billion on contraceptives in 2013. Roughly two-thirds of that was on contraceptive devices, including condoms, implants and intrauterine devices (IUDs, or 'coils'). Meanwhile, the IMS Institute for Healthcare Informatics estimates that in 2014 the world spent $100 billion on cancer drugs, and that figure has been growing at 6.5 per cent per year. Contraceptive drug expenditure is set to grow at just 1.3 per cent a year. Add to this the risk of getting sued, and the continued belief that men won't take a contraceptive pill, and Coelingh Bennink believes no drug company will get involved. 'This is a task for public organisations,' he says.

The WHO continues to fulfil this role – but it too has hit problems. In 2011, another progestogen–testosterone trial on over 200 couples, run by the WHO and the non-profit research organisation CONRAD, was stopped early. CONRAD announced two serious adverse events as the reason, although full details are still to be published.

Yet Anderson, who helped run the WHO–CONRAD trial, points out that some researchers are already offering the method to men outside of trials, and even using it on themselves. For him, the biggest obstacles are not scientific. 'Getting male fertility down to acceptable levels is difficult but not impossible, and there have been many years of experience of how to do that,' he says. 'What the field has really lacked is a champion with lots of money and enthusiasm. Thereafter you get industrial involvement.'

That champion may not yet have emerged, but in the USA, two women are at least providing the enthusiasm.

'We're talking about drugs men are going to take for a really long time, so the pathway for approval is long too,' says contraception researcher Diana Blithe. 'So when scientists say, 'I have a product in mice that looks promising, we'll have a drug in five years,' it's very unrealistic.' Nevertheless, she admits to being 'really excited' about the approaches she's supporting.

Blithe is director of the male contraceptive development programme at the US National Institute of Child Health and Human Development (NICHD) in Bethesda, Maryland. She's responsible for one of the largest pots of male pill research money available today and believes a hormonal method is most likely to do the job.

She points out that American men can already buy testosterone gels that could form part of a male contraceptive, and which show how to get a male hormone product approved. Adverts everywhere in the USA talk about 'low-T' – low testosterone levels – and the gels men can rub into their skin to treat them. Similarly, NICHD funds researchers at the University of California, Los Angeles and the University of Washington to do clinical trials using testosterone and progestogen in gels.

NICHD is also closing in on an elusive pill-form male hormonal contraceptive. Forms of testosterone that we can absorb from our stomach and gut rapidly break down in the body, meaning men would have to take pills three times a day. 'Would men take a pill?' Blithe asks. 'We think they will – but not every eight hours.' Therefore NICHD has developed a hormone that does the job of both progestogen and testosterone and only needs to be taken once a day. This too is moving into clinical trials.

Although she's enthusiastic about these ideas, Blithe stresses that NICHD can't do what Coelingh Bennink wants them to, and take male products through to approval independently. Instead, she and her colleagues are continuing to seek involvement from drug makers. 'Our hope is to show that it works well and men like it, and then a pharmaceutical company will recognise that it's safe,' she says. 'We are doing phase II now on the gels and if it works really well and we still don't have a partner, I don't know what the Institute's decision will be, whether they will want to continue.'

While scientists can work on how hormonal drugs are taken and their side-effects, one downside seems unavoidable. It takes one to four months to clear out already-made sperm and achieve the contraceptive effect, and a similar period for fertility to return. NICHD is therefore also backing research on non-hormonal methods that might be effective more quickly, but Blithe admits these are 'way further back' in animal testing.

If NICHD worked in the UK, they might therefore be interested in Nnaemeka Amobi from King's College London's non-hormonal 'instant male pill'. Also known as the 'dry orgasm pill', Amobi's contraceptive stops men releasing semen and the sperm it contains. He stresses that otherwise the normal physical processes involved in a man's orgasm are unaffected.

'The movement of semen from the testes to where it stays until you have the projectile phase of emission, called ejaculation, happens long before climax,' Amobi says. 'As soon as you're aroused, spermatic fluid is moved towards the seminal vesicles and prostate. Our pill stops that initial movement by inactivating the tubes that propel fluid from the testes to the prostate.'

Amobi and his fellow researchers started from two existing drugs that had caused dry orgasms as an undesirable side-effect. They redesigned the drugs to remove the original intended actions and focus on this. Animal tests suggest that they have succeeded. 'We used rams because rats and rabbits don't have seminal fluid like humans,' Amobi says. 'We tried boars, and boars produce 250 millilitres of semen. Can you imagine that? Rams have 1 millilitre, closer to humans' 2–5 millilitres.'

These tests show the method could become effective within 3–4 hours, and wear off after a day. 'A woman can say, 'Here's the pill – let me see you take it',' Amobi says. And as well as avoiding pregnancy, preventing semen emission should help reduce sexual transmission of semen-borne diseases, such as HIV.

One potential backer interested in the drug was the Parsemus Foundation, a small private organisation based in Berkeley, California. Ultimately, though, its founder Elaine Lissner faced a tough choice between funding Amobi's research and another promising new male contraceptive technology. She chose to spend the foundation's little cash on the latter. Amobi isn't bitter because, in his opinion, Lissner is the main reason people still talk about male contraception. But she still has regrets. 'It's shocking that they can't get backing for the first new idea about HIV transmission prevention in ages.'

Having started the Parsemus Foundation in 2005 with a little of her own money, Lissner has a personal relationship with how it's spent. In contrast to Blithe, she dislikes hormonal approaches because of their side-effects, and she also dislikes risk. Parsemus has therefore adopted an approach similar to one already tested in men, in India. But it's not a pill – it's a 'hydrogel' injected into the vas deferens, the tube linking the epididymis to the penis.

Called Vasalgel, it lets through semen but not sperm, and is intended to be washed out by another injection when men want the use of their sperm back. The blocked sperm are cleared from the epididymis and eaten up by immune cells, as happens normally if a man hasn't had an orgasm for a while. Lissner publicises Vasalgel energetically, and one glance at its thriving Facebook page should dispel any doubts that men would be interested. 'People are crazy for Vasalgel, desperate for it,' she says. 'We have over 32,000 people on the mailing list waiting to hear about clinical trials.'

One man who's keen to try it is Justin Terry, a married 30-year-old machinist who makes vehicle parts in Alabama. He and his wife don't have children, and his wife is taking the contraceptive pill. 'We've been married ten years,' Justin says. 'She doesn't want kids and neither do I, really. She wants to get off the pill.' The pill gives his wife tender breasts, and she is concerned about adverse effects of continuing to take it. As with hormonal approaches, his sperm would still flow for weeks after Vasalgel is injected, but this doesn't bother Justin. He has considered vasectomy, as have I, but has hesitated in part because it's not completely reversible. 'Vasalgel sounds like it will be reversible and would involve much less invasive surgery,' he says.

Parsemus's efforts have been helped by the David and Lucile Packard Foundation, also based in California, which provided $50,000 to help them test the approach in baboons. 'We expected to be out of money last year and we're not,' Lissner says. 'But the clinical trial is half a million dollars, so that's a different scale, and beyond that it's multimillions.' The trial will involve about 30 men and will test Vasalgel as a vasectomy alternative, without looking at reversibility.

Knowing the field's status, Lissner is not relying on government or the pharmaceutical industry. Instead, she's looking for backing from wealthy 'social investors' – and of course potential end users – and is publicising what might be possible in the field to bring interested parties together. 'The difference is that we have built an infrastructure where the public is able to channel its support,' she says.

§ § §

On Blithe's suggestion, I'm watching a documentary called The Great Sperm Race, made by the UK's Channel 4, showing the journey sperm make through a woman's uterus to her fallopian tube. It has cast people dressed in white clothing as sperm, dying in vast numbers. From millions of sperm ejaculated, just 20–100 get close enough to the egg to try to fertilise it.

As I watch, I imagine the white-clad actors instead represent the many possible male pills. There have been and still are masses of ideas, far more than I've been able to mention. Yet, like the unsuccessful sperm, so many have fallen by the way-side. I think of the contraceptive drug industry's current status and I can't help think we have missed its fertile period. If the perfect idea were to fight its way through development today, there's only a tiny chance that there would be a partner to meet it and eventually produce a fully formed male pill from it.

It seems obvious that if a new male contraceptive does make it to maturity, it will come thanks to the efforts of people like Blithe and Lissner. They, as much as anyone, are trying to create environments where the right technology can take seed. Without keen interest from the industry that we have traditionally relied on to supply our contraceptives, that requires enormous effort. Lissner's energetic exertions to concentrate support from men like Justin Terry, my dad and me could prove critical.

And Lissner is adamant that the ideas that seem to have faltered are not dead, they're just resting. 'We keep collecting new methods and never finish the ones we have,' she fumes. 'Pick one and make something! Finish the job!'

This story was first published on 12 July 2016
by Wellcome on mosaicscience.com

How virtual reality is aiding remote surgery

■ Jo Marchant

Mosso watches her carefully. A 54-year-old surgeon at Panamerican University in Mexico City, he's on a mission to bring virtual reality into the operating room, using the high-tech distraction technique to carry out surgeries that would normally require powerful painkillers and sedatives, with nothing more than local anaesthetic. He's trying to prove that reducing drug doses in this way not only slashes costs for Mexico's cash-strapped hospitals, but cuts complications and recovery times for patients, too.

But today, he's not sure if his headset is going to be enough. He hopes the virtual reality will help Ana to avoid unnecessary medication, but if she becomes anxious during the surgery, her already-high vital signs might spike. He has prepared an intravenous line, ready to administer emergency medication if required.

The surgeon pulls a large, pearly glob of tissue from Ana's thigh, his fingers easing under her skin as he carefully snips it free. Then he mops the blood and stitches the wound. The procedure has taken just 20 minutes, and there are smiles all round as Ana thanks the team. Because of the virtual reality, she says, she barely noticed the scalpel slicing her flesh: 'I was transported. Normally I'm very stressed, but now I feel so, so relaxed.'

The monitors back up her story. Throughout the surgery, her blood pressure actually fell.

§§§

In 2004, Mosso bought a Spider-Man game for his eldest son, and his life and career path changed. The game involved images projected onto a head-mounted display – an early form of virtual reality (VR). Mosso was struck by how immersed his son became in the game. 'His Mom called him to go to dinner and he didn't hear her, nothing. I thought, what if I use this on a patient?'

Mosso began using the game during upper gastrointestinal endoscopies, in which a flexible tube with a camera on the end is fed through a patient's throat into their stomach. The experience can be unpleasant and distressing. Patients often require sedation but Mosso encouraged them to play the Spider-Man game instead, to distract themselves.

He asked the patients to score their pain and anxiety during the procedure and in 2006 presented his results at the Medicine Meets Virtual Reality conference in California. The idea of using VR to reduce the distress of medical procedures was pioneered at the University of Seattle, Washington, where cognitive psychologist Hunter Hoffman and colleagues have developed a VR game called SnowWorld, to help patients endure wound care for severe burns. The researchers hoped that the illusion of being physically immersed in a three-dimensional computer-generated scene would move patients' attention away from their real-world pain. It worked: Hoffman's team has since shown in trials that SnowWorld reduces patients' pain during wound-care sessions by up to 50 per cent, as well as reducing pain-related brain activity.

But there has been relatively little work in other medical contexts. At the 2006 conference, Mosso met Albert 'Skip' Rizzo, a psychologist (and now director of medical VR) at the University of Southern California, who had been doing similar research with endoscopies. 'He presented 10 cases,' says Mosso. 'I presented 200.' Rizzo showed Mosso the expensive, state-of-the-art head-mounted displays he was using. 'It was another world,' says Mosso. But then Rizzo revealed the equipment with which he had begun – it was the exact same Spider-Man game.

'In this moment my life changed,' says Mosso. 'Skip saved me.' Impressed by Mosso's work, Rizzo donated a headset to him and persuaded a colleague, Brenda Wiederhold of the Virtual Reality Medical Center in San Diego, to let Mosso use some virtual worlds she had developed specifically for pain relief.

Mosso returned to Mexico with his new equipment and started to use VR in a much wider range of situations, from childbirth to recovery from heart surgery. It helped to relax patients across the board, but some of his most successful results were in minor day surgery, procedures such as removing lipomas, cysts and hernias, during which patients are awake but often sedated. He used a virtual scenario developed by Wiederhold called Enchanted Forest, in which users can explore rivers, lakes, trees and mountains. (The virtual world has to be relaxing, notes Mosso. A shoot-em-up game, no matter how distracting, might increase the risk of uncontrolled bleeding if the excitement raised patients' blood pressure.)

VR is now being studied by teams around the world to relieve pain in medical situations such as wound care and dentistry, as well as in chronic conditions such as phantom limb pain. But Mosso is still the only researcher to have published results on the use of VR during surgery. In one study of 140 patients he found that those using VR reported 24 per cent less pain and anxiety during surgery than a control group. He got similar results in a smaller randomised trial.

Offering patients VR also halved the amount of sedation they needed, and in many cases avoided its use completely. That represents an important cost saving for the clinics in which Mosso works; sedative drugs such as fentanyl and midazolam are 'very, very expensive', he says. He estimates that this reduced the cost of surgery by around 25 per cent, although he hasn't yet crunched the data to give an exact figure. Cutting drug doses should also reduce complication risks and recovery times for patients. Mosso is planning further trials to test this, but in general, he says, patients can go home an hour after surgery if they receive only local anaesthetic, whereas those who are sedated often need a whole day to recover.

'It cuts down on the cost, on the recovery time, and on the complications,' says Wiederhold. 'It's incredible. We still have not done that here in the US.' Gregorio Obrador, dean of medicine at Panamerican University, is impressed too. At first, 'I thought it was a little goofy,' he admits. 'I'm accustomed to giving pain medication.' But after looking at the literature on VR and pain relief, 'I'm convinced that it does work.'

Overall, Mosso has now carried out more than 350 surgeries using VR, and says he'd love to see it used as a routine component of pain relief in operating rooms. Offered alongside medication, he thinks the technology could transform how patients are treated during a wide range of procedures. But he has a bigger vision. What if VR could be more than an alternative to sedation during hospital surgeries? Could it help him to bring surgery to patients where sedation isn't possible, where there are no hospitals at all?

§§§

Mosso's Jeep Cherokee is full to bursting. Tents, plastic food boxes, surgical equipment, medication, sanitary products and bags filled with clothes, sweaters and shoes are squeezed into every available space inside and tied precariously to the roof. On the back seat are Mosso's wife, Veronica – a gynaecologist – their youngest son, Olivier, and, to keep the nine-year-old entertained, two baby iguanas recently captured from the forest near Acapulco, confined for the journey in a green net bag.

There's a long drive ahead. We are going to El Tepeyac, an isolated village hundreds of kilometres away in the mountains of Guerrero state. It's home to an indigenous Me'phaa community (often called Tlapaneco by outsiders), one of Mexico's poorest. 'They have been forgotten,' says Mosso. 'They live with cold, on top of the mountain. They don't have hospitals, clinics, nothing.'

As the high-rise blocks of Mexico City give way to sprawling shanty towns and then forested mountains, Mosso tells me about his father, Victorio. He was born close to El Tepeyac but left when he was 13, eventually becoming a teacher near Acapulco. He returned briefly to his childhood home after getting married, but never visited again until Mosso took him 40 years later. They found Victorio's youngest brother, Faustino. At first, neither brother recognised the other. 'They said 'You look too old!'' recalls Mosso. 'Then they were hugging, crying, a lot of emotions. It was the first time I saw my father cry.'

Mosso was shocked by the poverty he saw, with dwellings that he felt could barely be described as houses. The villagers asked him to examine a patient, an old woman with a fever who was lying in a puddle on the floor (there had been a recent flood, and it was the only place close to the fire). She had pneumonia; Mosso told them there was nothing he could do. 'She was my aunt,' he says. 'It was the last time I saw her. She died a few weeks later.' He pauses, eyes fixed on the road. 'That's why I go back. Because of my aunt.'

In 2000, Mosso and Veronica began to travel to El Tepeyac every few months. They helped the villagers to build and stock a basic medical clinic, and carried out simple surgeries. But a few years ago their trips stopped, due to a sharp rise in violence from the country's drug cartels. These organised criminal groups have been active across Mexico since the 1990s, producing heroin from poppies grown in the mountains here and exporting it to the US and Europe. Any violence was traditionally directed mostly at the authorities and each other, but since 2009 the cartels have increasingly targeted the general population with extortion and kidnappings.

The threat of violence is now routine for many Mexicans; the news here is filled with beheadings, mutilations and disappearances. On the freeway in the outskirts of Mexico City the day before, we had passed a group of four men, calmly crossing on foot between the busy traffic. One of them carried a young woman over his shoulder, either dead or unconscious, her dark hair spilling down past his hips. Mosso shrugged; for him the sight was nothing unusual. He works weekends at a hospital in this area and says he once had to order his surgical team to flee the operating room when a gunman entered the building, intent on killing their patient.

But the security situation is particularly bad in Guerrero, which is the country's most violent state, with one of the highest murder rates in the world.

According to a 2015 report by anthropologist Chris Kyle of the University of Alabama, Birmingham, illegal roadblocks, carjackings and kidnappings are routine here. The police have lost control, Kyle says, and there is 'near complete impunity' for the perpetrators. In 2009, Mosso and Veronica reluctantly decided that it was too dangerous to travel. 'We were coming to El Tepeyac four times a year,' he says. 'When the narc began, no more.'

But he's desperate to see his family, and worried about the health of the villagers. So although the security situation hasn't improved, he is now attempting the trip again. The obvious route from Mexico City is to take the highway via Guerrero's capital, Chilpancingo, to Tlapa de Comonfort, the nearest town to El Tepeyac. But the road from Chilpancingo to Tlapa – the main route for transporting opium out of the region – is 'hell', Mosso says, with many shootings and kidnappings. Instead we take a roundabout route through the states of Morelos and Puebla. We travel by daylight and eat on the move, making just one brief stop, in a deserted lay-by, during the nine-hour drive.

His caution pays off; the only sign of trouble is three cars travelling in convoy – 'When you see vehicles driving together like that, it's the narc,' Mosso notes as we pass – and once we reach the steep streets of Tlapa, he visibly relaxes. In this largely indigenous area, self-organised community police groups have been relatively successful in limiting the violence of the cartels. From Tlapa, the road gets higher and rougher as the sun sets, eventually becoming a narrow, winding track of mud and stones.

We arrive to find El Tepeyac in darkness; the only power line was recently blown down by a storm. The villagers line up to meet us with flashlights, wide eyes and smiles looming out of the black. The welcome is a little awkward – many of them don't speak Spanish, and Mosso doesn't speak Me'phaa – until they direct us to a long, plastic table beneath a high shelter and feed us chicken soup and tortillas, freshly cooked over a fire, with steaming lemon tea.

§§§

The sun rises to reveal the centre of El Tepeyac as a handful of brightly painted concrete buildings surrounding a covered basketball court, where communal meals and functions are held. Around 150 people live here, their homes scattered across the mountainside, each with space for vegetables, chickens and cows, and a large rain butt for fresh water.

There's a breathtaking view over slopes forested with pine and eucalyptus trees, with maize plants squeezed into every available space. (The terrain is also perfect for growing poppies, and although we don't see evidence of it in El Tepeyac, most communities in this region supplement their income in this way.) Mosso points out neighbouring villages – while most inhabitants of El Tepeyac are Me'phaa, the people in the next village belong to another indigenous group, the Mixteco, while the ones beyond that are Nahuatl, descendants of the

Aztecs. There's no cell or TV signal here and these communities have limited contact with the outside world; instead, they communicate with each other by two-way radio and closed-circuit television, all in local dialects.

Straight after breakfast, Mosso visits another of his aunts. She's small and squat with missing teeth and lives with her son and daughter-in-law in a mud-brick house with a roof made of corrugated iron. She holds her nephew and weeps. Her husband, Victorio's brother, has passed away since Mosso's last visit. Of ten siblings, only one is still alive.

Then it's time for work. We walk down a muddy track to a single-storey building with two rooms, bare concrete floors and shelves stacked with pills. 'We say it's a clinic,' says Mosso, 'but it's just a house.' Would-be patients – some are from El Tepeyac, others have walked from neighbouring villages – wait in an open porch while Mosso and Veronica set up tables and chairs inside. This morning, the two doctors will each hold an open clinic.

Mosso's first patient of the day is a young mother. Her seven-month-old baby, Hector, has a flattened forehead and plaintive cry. Mosso diagnoses microcephaly: the baby's brain hasn't developed properly. The Zika virus is causing cases of microcephaly across Central and South America, but Mosso doesn't think that's the case here; the mosquitoes that carry the virus don't usually live at this altitude (2,300 metres), and the woman says she hasn't visited the coast.

She shows no emotion as he explains her baby's condition, then she thanks him and leaves.

He gets through around 20 patients during the morning. One anxious man has red tracks on his thighs from the claws of a tarantula that crawled into his trousers while he was working in the fields. He has since developed sensitive skin and back pain, which he fears is due to the spider's poison. Mosso prescribes antibiotics for cases of parasitosis and kidney infection, and diagnoses tooth decay in almost everyone; there is little education here about oral hygiene. Diabetes is common, too, as the villagers routinely consume sugary drinks instead of water. Mosso lectures one patient after another: 'No Coca-Cola,' he says. 'Only one tortilla, not five.'

One old man comes in with a hernia untreated for 20 years. The nearest doctor is in Tlapa, explains Mosso, an hour's drive away but a difficult journey without a car. The government does subsidise medical care for indigenous groups, he says, but even when they are able to travel they are sometimes discriminated against – put off from treatment – or they simply don't know who to see or what care is available. Mosso writes several personal referrals to colleagues in Tlapa, which he hopes will accelerate the villagers' access to the care they need. He also identifies a handful of cases suitable for surgery here in El Tepeyac. But there's a problem – the village is still without power.

After lunch at Mosso's niece's house, which turns out to be perched on the mountainside up a muddy track so steep it makes the Jeep's wheels spin, the lights come back on; the surgery can go ahead after all. The clinic floor is briskly swept as Mosso and Veronica put on scrubs and lay out scalpels. A nine-year-old girl named Joanna is on a bed by the window, screaming for her mother. Mosso is going to remove a lump of cartilage from behind her ear. She is wearing jeans and a T-shirt, and has bare, dirty feet. Through the window children are playing, adults sit in chairs sharing home-brewed tequila, and the mountains stretch for miles. A fly crawls slowly over the paint-splattered floor.

Veronica fits the VR headset and the girl is immediately quiet. 'I see fishes,' she says. 'I see water.' Mosso has chosen for her an island world, with stone ruins and tropical fish beneath the sea. She remains still and calm until Mosso has finished stitching, then describes her experience. 'I have never seen the sea,' she says. 'I liked it. I felt that the water was warm.'

Then there are several lipomas to remove; these benign tumours are mostly harmless but if they cause pain, Mosso recommends surgery. He operates on a 54-year-old kindergarten teacher with two lipomas on her arm, and a man in his 20s who studied in Tlapa and has played video games before. The man is sceptical about the VR at first, but it was 'better than I thought it would be', he allows after the surgery.

Next is 31-year-old Oliveria, her dark, curly hair tamed in silver butterfly clips. She has four children, works as a farmer and has walked from a village one-and-a-half hours to the south. She has a lipoma deep in her back, which hurts when she moves. It is a slightly trickier case than the others but the lump is likely to keep growing, so Mosso thinks it's best to remove it now.

Oliveria lies on her front in black jeans and a bra as Veronica fits the headset; she's watching the same undersea world as Joanna. Mosso injects local anaesthetic into the lump, makes a cut, and his white-gloved finger disappears to the knuckle. He feels around. 'I'll have to open up the muscle,' he concludes. He extends the cut and pulls open the flesh with metal brackets before reaching deeper than before. Eventually, he manages to pull the fatty ball free. Veronica holds it tight with tweezers as Mosso snips around: success. But the undersea world is suddenly replaced by an error message. The laptop wasn't plugged in, and the battery is about to fail. A few seconds later, Mosso and Veronica realise that Oliveria has lost consciousness.

Everyone's moving. They turn the patient onto her back, Mosso rubbing her chest and shouting 'Vamos a la casa!' while Veronica waves alcohol-soaked cotton wool under her nose. The pain triggered Oliveria's blood pressure to drop suddenly, explains Mosso, causing her to faint. He inserts an intravenous line with fluid to restore her blood pressure. Shortly afterwards Oliveria moans, and bats away the cotton wool. 'Breathe slowly,' instructs Veronica. Mosso swats a fly from her face.

After a few minutes, they roll Oliveria onto her side to sew up the wound. Mosso doesn't have the facilities here to sedate her, or offer her any painkillers more powerful than the local anaesthetic, so he plugs in the laptop and switches the VR back on. Veronica keeps Oliveria talking as Mosso works. 'What do you see?' she asks. 'Fishes, water, stones,' comes the reply. Then they help her to her feet and walk her to a bed in the next room. There's no stand or hook for the IV line so after some searching Oliveria ties it to an old floor lamp, which he balances on a table by the bed, next to Olivier's iguanas, happily munching lettuce on a plate.

'It looks easy, but we never know at what moment we can have a surprise,' says Mosso when the crisis is over. 'In a hospital I'm relaxed, because the monitor tells me the patient's heart rate, breathing, blood oxygen. There's an anaesthesiologist, scrub nurse, other surgeons. But here, we're far away from the hospital and my colleagues. With or without surprises, I'm worried. What if something happens here and I don't have solutions? Tlapa is far away.'

Does he think it's worth the risks? 'Yes,' he says without hesitation. 'They don't have the opportunity for surgery otherwise. And the risk is very low. In 350 patients, I just had one with this complication.'

Half an hour later, Oliveria is ready to leave. 'I didn't know I was going to have surgery today,' she tells Mosso and Veronica. 'Thank you.' Mosso gives her paracetamol and antibiotics, and instructs her to take a taxi home. She has asked to keep the lipoma so he hands her the twisted, blood-stained lobes in a small pot of alcohol. Her hands are shaking as she takes it.

§§§

Next morning there's an impromptu farewell party on the basketball court. The village brass band accompanies a range of traditional Mexican dances, including one in which Mosso does a surprisingly athletic impression of an iguana.

He wants to leave early – today we will drive to Acapulco, where he plans to visit family (and release the real iguanas), before returning to Mexico City. Despite the circuitous route he has planned, it is unwise to be on the roads around Acapulco after nightfall. But there's another line of people at the clinic. Veronica hands out the clothes and supplies from the Jeep – donations from Olivier's school – while Mosso sees the patients. There's one more case for surgery: a boy with a haemangioma (a benign tumour of blood vessels) on his head. There isn't a strong medical need to remove it, but the boy is being bullied by his friends – 'they say it's an insect,' translates Mosso – and his mother is desperate.

Mosso agrees to the surgery, but once that's done more patients arrive – they've walked an hour to see him. Mosso says no. It's already early afternoon, we have to go. We drive seven hours without stopping, the air ever warmer as we leave the mountains and climb down towards the sea. He's agitated, pushing 90 miles an hour along the long, straight coastal road, but we lose the race. The sun sets and we speed towards the city in darkness. Then cars coming the other way begin to flash their headlights, and shortly afterwards we're waved to a halt by a group of armed men in military attire.

Mosso knows the drill. Quickly he opens his window, flips on the interior light and calls his son into the front. They're looking for enemies, he says. As long as they can see we're not hiding anything, they should let us through. Sure enough, the gunman looks inside and waves us on.

Once at his home in Acapulco, in a gated apartment complex, Mosso reflects on the trip. Apart from the fainting episode the patients all did well, and we travelled safely. 'It was successful,' he says. 'I'm happy with the results.'

He has collected data on all of the surgeries he carried out, and hopes that his experiences will encourage the use of VR to help patients in other under-resourced communities around the world. The cost of VR headsets has been prohibitive, but in the last year or two, the release of cheap devices such as the Samsung Gear VR (which costs less than £100) and even the Google Cardboard (£3), as well as the growing number of virtual worlds freely available online, have transformed access to the technology. 'When we started, virtual reality was expensive, difficult to get and difficult to set up,' says Mosso. 'Today, everyone can use it.' Although Mosso connected his headset to a laptop in El Tepeyac, he has previously shown that the technique works just as well running from a mobile phone, perfect for relieving pain in difficult locations. 'There's no heavy equipment,' he says. 'It's very easy to use.'

Meanwhile he is already making plans to return to El Tepeyac. During our trip, he met with a local government representative who wants him to visit not just that village but neighbouring indigenous communities too. That would take time and money that Mosso doesn't have, but he's trying to convince some of his colleagues in Mexico City to help, and hopes that soon he'll be able to return to Guerrero with a team of surgeons, perhaps in spring 2017. He dreams of one day reaching even more remote communities – villages deep in the mountains that he has heard about but never visited, that have practically no contact with the outside world.

Mosso is one of the most upbeat people I have met. Tonight, though, his optimism is tempered. He says his overwhelming emotion on leaving El Tepeyac was anger. 'I've seen some economic development,' he says. 'But my family are living in the same house, they are wearing the same clothes. All I gave is nothing. When I said goodbye I felt angry with myself, because I can't do anything for them.'

He's painfully aware that it will take more than VR and donated sweaters to solve the problems of the people of El Tepeyac – and his country. But he's working to help them in the only way he knows.

This story was first published on 31 January 2017
by Wellcome on mosaicscience.com

Shhh! But what exactly is the menopause?

■ Rose George

My physio, a young woman called Lucy, was simply making conversation. She wanted to distract me from the serious discomfort she was about to inflict by massaging the nerves around my painful posterior tibial tendon, an ankle injury I assumed I had brought on by running too much. 'My mother's post tib has ruptured,' she said. 'It's really common in menopausal women.' This definitely worked as a distraction. What did all this have to do with the menopause, I asked? She looked surprised, because to her the answer was obvious: 'Collagen.'

Suddenly something clicked, and not just in my ankle. For about a year, the skin on my hands had been peeling, monthly. I'd seen GPs and pharmacists and been given various remedies, from 'try thick hand cream' to 'drink more water'. Lucy's comment made me research more: oestrogen is related to collagen production, and when oestrogen levels start to change in women who are in the stage approaching the menopause (the so-called perimenopause), all sorts can happen.

Perhaps I should have known this. I've already had one menopause. It was chemically induced as a treatment for my endometriosis, a condition where the cells that line the inside of the uterus (the endometrium) grow elsewhere in the body. I was given a course of injections of leuprorelin, a drug that blocks the production of oestrogen. Leuprorelin is not fussy: it can block testosterone in men and oestrogen in women, hence it's used to treat prostate cancer and chemically castrate paedophiles, as well as to calm down inflamed female pelvises.

'You may have some vasomotor symptoms,' said my consultant, adding that the whole thing wouldn't last more than six months. He was right about the first and wrong about the second. 'Vasomotor' refers to the constriction or dilation of blood vessels. In the case of the menopause, the results are hot flushes and night sweats. I remember sitting at posh dinners, pouring sweat and being thankful my dress was black. I carried a fan and deodorant at all times. I stank. My moods sank to alarming depths. I stopped sleeping.

I longed for it to stop. It did, once I finally went on hormone replacement therapy (HRT). I'm now 'naturally' in the perimenopause, the stage before the menopause that can last several years. But even though I'm postmenopausal and perimenopausal all at once, Lucy's comment made me realise that I still knew too little about hormones and the menopause. And in this, I was completely normal.

§§§

A few things science doesn't know about the menopause: what it's for, how it works and how best to treat it.

As a comprehensive review in *Nature* put it, 'the functional lifespan of human ovaries is determined by a complex and yet largely unidentified set of genetic, hormonal and environmental factors'. Also poorly understood is what happens when the ovaries begin to fail and hormone levels begin to fluctuate.

Perhaps we should be sympathetic to this ignorance. The menopause doesn't make much sense, biologically or intellectually. Humans are one of only three animal groups that we know of to experience it (the others are killer whales and short-finned pilot whales). As one recent book on primate ecology puts it: 'Menopause is still considered a distinctly human trait.' That we live so far beyond our reproductive usefulness is a puzzle that was answered, supposedly, by the 'grandmother hypothesis'. By this reasoning, human females live beyond their reproductive years because their presence benefits their children and grandchildren. One aspect of this relates to the fact that humans are no longer well-designed to give birth, because walking upright and having a large brain has led to a pelvis size which makes giving birth standing up or without help extremely difficult. Older females, then, can be useful even when they are not producing offspring.

This idea was questioned by the ecologist and biologist Craig Packer, though by studying lions and baboons, not women. He found that the presence of post-reproductive females gave the animals no particular advantage: young lions and baboons with grandmothers fared as well as ones without. Another theory, that the menopause is a product of our longer modern lifespans (that we died so young, it didn't have chance to exist), is easily skewered. There are plenty of accounts of women living to a good age throughout history. The concept of the menopause, though, is modern: the phrase was invented in 1821, but it was only in the 20th century that the concept became dogma.

At least medical definitions seem clear on what the menopause is: a biological stage in a woman's life marked by the cessation of periods because of a reduction in the function of the ovaries. This seemingly straightforward statement conceals vast and mysterious depths. A female human is born with over a million eggs in her ovaries. Every month, she releases one, a process triggered by the release of hormones, including oestrogen. After the age of 40, the ovaries begin to secrete less oestrogen, and, in the wonderful understatement of the charity Women's Health Concern, this causes 'the body to behave differently'.

Oestrogen is involved in a range of bodily functions, and oestrogen receptors are found in cells throughout the body: the brain, the breasts, the bones, the belly. The hormonal fluctuations of the perimenopause and menopause are most famously involved in creating hot flushes, but they may also be linked to cognitive impairment ('brain fog'), irritable bowel, nausea, aching joints, cracking or peeling skin, depression, vaginal atrophy and dryness, lowered libido, memory loss and sleep disorder, osteoporosis, and flat feet. This is not an exhaustive list, and some of it lies in the realm of internet forums and anecdote. But I can vouch for most of it.

Every woman's body reacts differently to changes in oestrogen levels, making a certain diagnosis of the menopause difficult. New guidelines from the National Institute for Health and Care Excellence (NICE) – an authoritative body that provides national guidance and advice to improve health and social care in England and Wales – warn against anything but observation for diagnosis: even the usual tactic of testing a woman's follicle-stimulating hormone (FSH) level is pointless, given how much FSH can fluctuate. In women aged 40–45, FSH tests can be done, but for nearly everyone else, the safest marker of the menopause is absence: the 'lost ovarian function' assumed if a woman hasn't had a period for 12 months. The perimenopause, which I'm in, can be diagnosed by erratic periods, perhaps vasomotor symptoms, and – this is purely a personal definition – the sense that things aren't quite right. That you are, biologically speaking, losing it.

§§§

Vaginal lubricant? Yes please. Physicool spray to calm down hot flushes? God, yes, even though my free tote bag is beginning to bulge. Isoflavones that supposedly balance oestrogen upheavals? OK, I'll have a few boxes. Why not? I'm already taking magnesium for the cognitive fog, vitamin D and the antidepressant citalopram for low moods, a herbal potion containing black cohosh and rhodiola for overall calming and balance, plus a multivitamin for good luck.

I was acting like a kid in a sweetshop, but I was in fact in the serious surroundings of the exhibitors' room at the British Menopause Society Annual Conference, held in a conference centre near Swindon. Unused to the glare or even interest of the press, the British Menopause Society treated my request to attend with puzzlement before accepting. I had already attended the morning session in a hall packed with doctors, nurses and therapists and listened to presentations on premature ovarian failure, new and better drugs for endometriosis, and the risks of pulmonary embolism (changes in oestrogen levels can affect how the blood clots). There was a short yoga demonstration by delegates from the Indian Menopause Society – featuring the slogan 'Add years to your life and life to your years' – during which suave consultants from Harley Street did sun salutations along with the rest of us.

I was there as a journalist, but also as a perimenopausal woman preparing herself. I took all the freebies I could because I want to be equipped for what's ahead. I learned a lot from the speakers: that cardiovascular disease is the most common cause of death in women, outstripping breast cancer tenfold (according to cardiologist Peter Collins), and that women can be prescribed remedies thoughtlessly and crassly. Psychosexual therapist Trudy Hannington described a woman who had been given a big tube of vaginal lubricant for dryness and an equally big recommended dose. 'She followed the instruction,' Hannington told the audience, 'and reported back that she was squeaking.'

The overwhelming message was consistent: a condition that personally affects half the population is woefully neglected. There is neither enough data nor enough drugs. The lack of attention paid to the menopause, and to women's health in general, has always made life difficult for anyone trying to care for menopausal women. In the early 2000s it became much harder.

§§§

In 2002, women who approached the medical profession for help with menopausal troubles were routinely prescribed HRT. The standard formulation for women who still had a uterus was a combination of oestrogen and a progestogen: either progesterone (derived from plants) or progestins (synthetic progestational agents which act like progesterone). The oestrogen is to replace the body's falling levels and the progestogens to protect the endometrium: though the mechanism is unclear, adding oestrogen without a progestational agent increases the risk of endometrial cancer.

In the USA, the most common HRT was a blend of conjugated oestrogens sold under the brand name Premarin, short for pregnant mares' urine because it was derived from the urine of captive horses in North Dakota and Western Canada. By the mid-1970s, it was the fifth most prescribed drug in the country, and it's still the one of the largest-selling, most commercial HRT products in the USA. According to sales figures, in 2014 it was the 38th most prescribed branded drug in the USA.

Then the results of the Women's Health Initiative were published. The Initiative was a programme of research launched in 1991 throughout the USA. Between 1993 and 1998, 27,437 women aged 50 to 79 enrolled in the Initiative's hormone study. Of these, 16,608 women who had an intact uterus were in the study of oestrogen plus progestin and 10,739 without a uterus participated in the trial of oestrogen alone.

Compared to a placebo, the oestrogen and progestin HRT was shown to cause 'increased risk of heart attack, increased risk of stroke, increased risk of blood clots, increased risk of breast cancer, reduced risk of colorectal cancer, fewer fractures [and] no protection against mild cognitive impairment and increased risk of dementia'. The relative risk of getting breast cancer was given as 26 per cent. The results were so shocking that the study was stopped in 2002.

The press headlines were loud, immediate and everywhere. The *Daily Mail*, in 2002: 'HRT linked to breast cancer'. The *Guardian*: 'HRT study cancelled over cancer and stroke fears'. Some articles were better than others, but the worst ignored the fact that the oestrogen-only HRT study was continuing. They also failed to distinguish between relative risk – the risk posed to the study group of women being given oestrogen and progestin relative to the risk posed to those being given a placebo – and excess risk, the actual increase in risk between the two groups. In fact, as the Women's Health Initiative researchers wrote in the *Journal of the American Medical Association*, in terms of breast cancer and stroke, the excess risk was just eight more strokes and eight more invasive breast cancers per 10,000 person-years.

The results of the UK-based Million Women Study, published in 2003, added to the alarm. Led by Oxford professor Dame Valerie Beral and funded partly by Cancer Research UK, the results seemed to show that breast cancer risk was doubled in women taking HRT. The study ascribed 20,000 cases of breast cancer per decade to HRT use, with 15,000 of those related to oestrogen-progestogen use.

In August 2003, the UK's Committee on Safety of Medicines circulated a letter to GPs and other health professionals telling them that long-term use of oestrogen and progestogen HRT was associated with 'an increased incidence' of breast cancer. Although it recommended that 'the results of the Million Women Study do not necessitate any urgent changes to women's treatment', it also said, in an accompanying patient information leaflet, that 'the longer HRT is used, the higher the risk of breast cancer'.

The effect of all this was profound. 'Everyone stopped prescribing,' says Julie Ayres, a doctor who runs a menopause clinic in Leeds, England. 'They don't have time to read beyond headlines.' Although a circular from the Committee on Safety of Medicines later that year repeated that short-term HRT was favourable for menopausal symptoms, HRT prescriptions still dropped by about 50 per cent in the UK between 2002 and 2006. In the USA, prescriptions of the two most common HRT brands, Premarin and Prempro, dropped from 61 million in 2001 to 21m in 2004. Newspaper headlines bombarded women with the message that HRT was dangerous.

The bombarding must have worked: even when I was in great distress with my chemical menopause, losing days of work to insomnia and hot flushes, struggling with depression and not far from a breakdown, I resisted it. Somewhere in my head I thought 'breast cancer'. When I eventually did take HRT, after I couldn't stand the insomnia any more, it was magic. I could sleep and think straight again. But I still came off it as quickly as I could.

§§§

In 1948, the obstetrician Dame Josephine Barnes gave a series of talks on women's health on BBC radio covering ovaries, bleeding and hormonal changes. There was uproar. The head of the Home Service, wrote Jenni Murray in the *Guardian*, 'spluttered [that] 'the inclusion of such a talk represents a lowering of broadcasting standards. It is acutely embarrassing to hear about hot flushes, diseases of the ovary and the possibility of womb removal transmitted... at two o'clock in the afternoon.' Nearly 70 years on, one of the few safe places to talk about menopausal women is in humour, and not always the gentle type:

Comedian Jeff Allen: '*My wife started the menopause. There are days when I lie in bed and dream of the good old days of PMS.*' Or, '*I tell my boys, Mom's going through some stuff. The nights when you don't do your homework and she gets mad and yells at you, it's going to be a little different now. She might start crying and stab you.*'

A classic mum joke: '*Your mum's so stupid, she thinks 'menopause' is a button on her iPad.*'

Joan Rivers: '*Had a friend going through menopause come to lunch today. Her hot flash was so bad, it steam-cleaned my carpet.*'

According to a video interview on the excellent website healthtalk.org, a woman named Maria, who used to work on a supermarket check-out, felt she could do nothing but join in when male colleagues laughed at her sweats. 'You get your blonde jokes, you get your menopause jokes,' she says. There is also empowering humour on websites, fridge magnets and tea towels. 'I don't have hot flushes. I have short, private vacations in the tropics.' Or, 'Real women don't have hot flushes, they have power surges.' There is a successful feel-good menopause show called *Menopause the Musical* (including the number 'Stay-in' Awake/Night Sweatin'') and plenty of blogs and sites urging women to embrace this positive change. I'm glad of all of it, though I'm not sure calling a hot flush a power surge is going to make them less distressing or smelly.

We can thank the French for at least having a word for this peculiar stage in a woman's life. 'Menopause' comes from *ménèspausie*, which in turn comes from Latin via Greek (*mens*, a month, and *pausis*, a pause) and simply means a cessation of the menses. I prefer the word 'climacteric', which is still used by medical professionals (and the title of one of the few dedicated journals on the menopause). Climacteric comes from the Greek for 'rung of a ladder' and means a critical stage or turning point. I like the dramatic sound of it, because, having had one menopause already, I know that it can feel dramatic: tragic and comic all at once. The word 'oestrogen', meanwhile, is derived from *oestrus*, a Greek word mostly translated as 'gadfly' or 'frenzy' (but sometimes as 'verve') and the suffix 'gen' ('producer of').

The biological fact of the menopause pre-dates this vocabulary. As Louise Foxcroft wrote in *Hot Flushes, Cold Science: A history of the modern menopause*, Aristotle, Galen and others knew that a woman stopped bleeding and lost her ability to reproduce. This change was thought to start at 50, though several sources, including the personal physician of Justinian I, state clearly that it can begin as early as 35, especially in those who are 'very fat'.

Foxcroft's history is a jolly escapade through the dreadful attempts by mostly male medical professionals to deal with the peculiar creature that is a woman who has lost her reproductive capacity and therefore – supposedly – her usefulness. The Victorian surgeon Lawson Tait thought that the solution to 'climacteric discomfort' was to lock women up. Mental illness was widely attributed to 'uterine trouble'. Throughout history, postmenopausal women have been variously considered sexless, shrewish, whorish, dangerous, hysterical and pointless.

§§§

At the British Menopause Society there was deep frustration about the impact of the Women's Health Initiative trials. Plenty of studies since have persuasively punctured the Initiative's findings – that HRT causes breast cancer – but have received little publicity. A special issue of *Climacteric* in 2012 re-examined the trials and their reception ten years on. Although lead author Robert Langer calls the trials 'sound', there were problems: the average age of actual participants was 63, yet the findings were initially presented as pertaining to all menopausal women. A statement attributed to the then Acting Director of the Initiative Jacques Rossouw said that 'the adverse effects of estrogen plus progestin applied to all women, irrespective of age, ethnicity, or prior disease status'.

A paper released by the Women's Health Initiative (WHI) authors in 2013 repeated the message that had been lost in the breast cancer furore: that HRT is useful for managing the symptoms of some (probably younger) women, but that the 'WHI trials do not support the use of this therapy for chronic disease prevention'. Lead author Rossouw, who works for the Initiative's sponsor, the National Heart, Blood, and Lung Institute, said: 'While the risk versus benefits profile for estrogen alone is positive for younger women, it's important to note that these data only pertain to the short-term use of hormone therapy.' In fact, wrote Langer in *Climacteric*, 'the WHI deserves credit for evaluating, and ultimately halting, what had become an increasingly common clinical practice of prescribing menopausal hormone replacement therapy (HRT) for women well past menopause or at high risk of coronary heart disease, with the expectation of providing cardioprotection'.

If the study were published afresh, the British Menopause Society wrote in a press release last year, 'there would be far less impact on postmenopausal women today'. It would be widely understood that prescribing HRT to perimenopausal, menopausal or recently postmenopausal women is far different to prescribing it to women ten years into the menopause.

Epidemiologist Samuel Shapiro was the lead author on a series of articles published in 2011 that questioned the methods of both the Million Women Study (MWS) and the Women's Health Initiative. A ‹properly designed cohort study›, Shapiro and colleagues wrote in their article about the MWS, should have excluded breast cancers already present at the start of the study. In conclusion, they wrote: 'HRT may or may not increase the risk of breast cancer, but the MWS did not establish that it does.' The reaction of Valerie Beral, the lead researcher on the Million Women Study, was unequivocal, claiming that their review of it was a 'restatement of views held by many consultants to HRT manufacturers (as these authors are) attempting to dispute evidence about the adverse effects of HRT'.

Shapiro and his coauthors say that their critiques were not funded by the pharmaceutical industry and were independent. The footnotes of their review of the Million Women Study confirm that the review was not commissioned and was peer-reviewed. The paper also says that all of the authors had consulted in the past with manufacturers of products discussed in the article (and that all but one were doing so at the time of publication). It's not uncommon for researchers working in this field to have conflicts of interest, such as lecturing on behalf of and consulting for HRT manufacturers.

When I reached her by phone, Beral wouldn't comment on seismic changes in menopause research, such as the recently published NICE guidelines. 'I haven't read them.' But, in late 2015, when the media leapt on a small piece of unpublished research presented at a conference with headlines such as 'Ignore health scares, HRT is safe, say scientists', Beral said on the *Today* programme and elsewhere what she said to me: 'The effects of HRT have been extraordinarily well-studied. We do understand them very well. We know the effects on the ovaries, breasts, [of] thrombosis. We know that the risks start as soon as you start taking it. There's little doubt about it. People shouldn't use words like 'safe'; women should be explained what the risks are.' (The research presented at the conference was not about cancer, said its author, Lila Nachtigall, who described British press coverage of it as 'ridiculous'.)

§§§

Where are we now? Go to the website of Cancer Research UK and you will be told: 'The evidence that HRT can cause some types of cancer (breast, womb and ovarian) is strong.' Go to the

British Menopause Society website and its fact sheets will tell you the risk of cancer is 'small' (breast) or 'not high in statistical terms' (ovarian). Go to your GP and anything could happen.

Hannah Short, a trainee GP, and Natasha North, convener of Menopause UK, launched the #ChangeTheChange campaign in March 2015 in frustration at the confusing, poor information available not only to women, but to medical professionals. 'The menopause wasn't in any of my textbooks,' Short told me during the British Menopause Society conference coffee break. She's heard of women going to one GP to be put on HRT, then going to another who takes them off it. She's heard of one GP who said that women just need to pull themselves together. She told me of a nurse who had gone through a surgical menopause who was treated as a hypochondriac when she complained her oestradiol treatment wasn't working.

Most patients who end up in Julie Ayres' menopause clinic in Leeds arrive with preconceptions. 'They say, 'I know there's a risk of breast cancer." But they're so desperate, they come anyway. 'They come with palpitations, anxiety and panic attacks and think they're going crazy.' They're not, but they are suffering from the wide-ranging power of oestrogen in the body. 'As soon as they say they're having palpitations,' says Ayres, 'the GP won't prescribe HRT because of the cardiac risk.'

This would infuriate some speakers at the British Menopause Society conference, where John Stevenson, a consultant metabolic physician at the Royal Brompton Hospital, presented research on the protective role that HRT can have on the heart. He is so convinced of the benefits he's prepared to prescribe HRT, because, according to him, it is 'probably the best treatment for postmenopausal women, [though] sadly only one cardiologist seems to know this... If women come to see me who are at risk,

I ask them if they've had a hot flush so I can prescribe HRT,' he says. 'If they say no, we turn the heating up.'

It's a good joke, but he is deadly serious: 'There is hard evidence of the protective effect of oestrogen for adverse cardiac events. There's no firm proof that HRT causes breast cancer.' He is dismissive of the Women's Health Initiative study (and was one of Samuel Shapiro's coauthors on the series of critiques published in 2011). 'They got the same dose of hormones no matter what age. Great for a 50-year-old, absolute poison for a 70-year-old. No-one in this room would do that.'

The *Daily Mail*, which recently published a powerful and useful series on the menopause, often runs articles about 'bioidentical hormones', also known as bespoke HRT. Yehudi Gordon runs a bioidentical hormone clinic in Harley Street. He is slender, tanned and looks 20 years younger than his 73 years, and he is evangelical about the benefits of bioidentical hormones.

They are better, he said when we met over coffee near his clinic, because the oestrogens are derived from plants such as yam and soy, and the progesterone is micronised (finely ground). Both these facts, he claims, mean bioidentical hormones are better processed by the human body than conventional preparations. He gives me a handout which explains further: the molecular structure of Premarin, it reads, 'may bear some similarity to that of human hormones, [but] it has been altered'. Other 'branded and patented HRT consists of synthetic hormones that have a different molecular profile to those produced in the body'.

With his bioidentical therapy, patients have blood taken and are prescribed a particular hormone combination according to their hormonal levels, which is made by a compounding pharmacy (one that can make up its own preparations). To listen to Gordon, you'd think he had found the holy grail.

It's persuasive. I leave almost tempted to make an appointment, despite the hefty private fees and cost of treatment (though Gordon says the HRT, daily, costs little more than a cappuccino). But other menopause specialists are circumspect. The bespoke preparations are prepared by a compounding pharmacy, but as Heather Currie wrote in an issue of *Menopause Matters*, 'there are currently no controls or regulations on the production, prescribing or dosing of bioidentical hormones'. In the USA, custom-compounded hormones, as they are known, are not regulated by the Food and Drug Administration.

'Bioidentical is just a brand,' says Nick Panay, a leading gynaecologist. 'We can tailor HRT too: it's the same stuff.' Julie Ayres has tested oestrogen levels in women taking 'bespoke' hormones and found them to be far too high. 'We can try combinations of oestrogens and progesterone,' says Ayres. 'We can prescribe bioidenticals. There are so many types of HRT. It's great when we get it right. Women tell me they've got their life back. And I can't tell you how often I've heard, 'Thank you for taking me seriously.''

§§§

I noticed a few months ago that my brain now hesitates, very slightly, when asked to choose between left and right. I'm dropping things more and being clumsy. For a whole day recently I was convinced that December followed October and was genuinely disturbed when I realised it didn't. I wrote a blog post recently that details some of these occurrences. Current ailments: jaw pain, dry eyes that make me feel like my eyeball is actually a hedgehog, poor sleep, constant tiredness. All can be linked to hormonal changes in my body. But am I ill? With all this talk of symptoms, you'd be forgiven for thinking so. Writers like Louise Foxcroft and Roy Porter have queried the medicalisation of something that is a natural and inevitable stage in women's lives. 'Modern attitudes to the menopause,' wrote Foxcroft, 'arise directly out of a poisonous history of lack and loss, disease and decay.' This view sees the menopause as just another biological stage, no more alarming than any other. Yet this position would be questioned by many clinicians and professionals working in the field, and many women in the midst of this life stage.

What's a menopausal woman to do? Perhaps, she must be patient, and become a patient. NICE is deciding which treatments are to be available on the NHS. In November it published official NHS clinical guidelines on the menopause for the first time ever. ('If you had a condition that affected all men,' says Heather Currie, 'it would be taken more seriously.')

The guidelines, which were out for consultation for six weeks, are both ground-breaking and cautious. They say that HRT 'is a highly successful treatment for common symptoms of menopause', and that HRT with oestrogen alone 'is associated with little or no change in the risk of breast cancer'. They add that oestrogen and progestogen can be associated with an increase in the risk of breast cancer, something that is acknowledged even by supporters of HRT: the ability of progestins to disrupt cell growth, though the mechanism is unclear, has long been known. Micronised progesterone, where the particles are smaller, is better tolerated than synthetic progestins, and has fewer side-effects.

When I took HRT it worked wonders for my hot flushes, but devastated my libido. Some practitioners think testosterone can help with lowered libido in women, although evidence is lacking. A transdermal testosterone patch aimed at women – brand name Intrinsa – was taken off the market in 2012, as were testosterone implants shortly after.

The NICE guidelines suggest that testosterone supplementation can be considered for menopausal women with 'low sexual desire' if HRT alone is not doing the job. A footnote adds that as testosterone doesn't yet have a UK marketing authorisation for this use, the prescriber 'should follow relevant professional guidance'. In essence, as the psychosexual therapist Trudy Hannington made clear at the British Menopause Society conference, this means prescribing male-specific products judiciously. 'We use a tenth of the male dose. One gynaecologist prescribed a whole tube a day and wondered why the woman developed black hairs and was jumping [up to] the ceiling.' Hannah Short has heard other doctors discussing female patients who have come to them. 'They were so dismissive of normal symptoms.' One woman who asked for testosterone was dismissed with, 'She just wants a sex drive.' Of course she did, and what's wrong with that?

§§§

The menopause is not monolithic. Reactions to it can vary widely across cultures and geography, and according to diet, lifestyle and fitness, as well as age. While UK websites cite that around 75 per cent of menopausal women report having hot flushes, the number of Japanese women having them is reported variously as low as one in ten or one in eight. Yet when Margaret Rees, a gynaecologist and the Editor-in-Chief of *Maturitas* ('An international journal of midlife health and beyond') visited Japan, women told her they have flushes, they just don't talk about them. And there is cultural baggage around the menopause that can distort matters: while some depression is related to hormonal upheaval, some may be due to the disparaged position menopausal women believe they are in. In Rajput culture in India, wrote Foxcroft, the menopause can be seen as liberating, as women can remove their veils and mix more widely, including with men.

There is no doubt, however, that the population of women suffering symptoms is huge and under-served. You can glimpse this in certain studies, such as one from the Trade Union Congress which found that 45 per cent of safety representatives interviewed said their managers didn't recognise problems associated with the menopause. A study by Nuffield Health found that 72 per cent of women felt unsupported at work when menopausal, and that 10 per cent of women considered leaving their jobs as a result. A study by the University of Nottingham released in 2011 reported that nearly half of women found it difficult to cope with the menopause at work. Nearly a fifth thought it affected how their colleagues and managers perceived their competence.

According to Menopause UK, there are only 29 menopause clinics in the UK to serve the 13m women – a third of the female adult population – who have reached the menopause, are currently going through it or are postmenopausal (and may have ongoing symptoms). Coverage is inconsistent: of course, not every menopausal woman needs treatment, and still less a specialist clinic, but even so, the coverage is illogical. The North of England has two clinics for 2.5m women: the

NHS in the Midlands and the East of England has seven. Most menopausal women go to their GP first, if they seek help at all. One retrospective study published in 2010 found that 18 per cent of women aged 45–64 consulted their GP for menopausal symptoms at least once throughout 1996. By 2005 this had dropped to 10 per cent. A 2012 study found that 60 per cent of women cope with their symptoms without any contact with healthcare professionals, preferring to get advice from friends, family and the internet. Yet 10 per cent live with symptoms for up to 12 years.

When I asked the people I interviewed what the most exciting research on the menopause is, they struggled to answer. Some, though, were hopeful of a new combined HRT drug that contains equine oestrogen and bazedoxifene, a selective oestrogen receptor modulator, that can modulate any damaging effects of oestrogen on uterus and breast tissue. 'All useful studies were stopped in 2002' is Julie Ayres' take on it. The menopause, says Margaret Rees, 'is not a disease, but it is an opportunity to address other issues in women's health'. Not just bones and breasts, either. The NICE guidelines advise practitioners to 'explain to menopausal women that the likelihood of HRT affecting their risk of dementia is still unknown'.

I don't think the menopause is a disease either, but it's already affecting my health – you try writing about oestrogen receptors, endocrine pathways and endometrial cell division while battling perimenopausal brain fog. And if preparing for it requires medicalising it, then I will. Maybe. For now I'm preparing by running, keeping strong and eating well. I haven't yet given up coffee or alcohol, but that may change when the hot flushes begin again. And I will be heading for the menopause clinic as soon as my periods have stopped for good, perhaps before. But

I'm still not sure if I'll take HRT. I want to protect my bones and heart, but the residual fear of cancer is still too deep, however debunked. I'd like to end on a positive note, one of clarity and conviction. But instead I'm just confused. And if that's the case for me, after months of reading, research and talking to experts, what chance of understanding does anyone else have?

This story was first published on 15 December 2015 by Wellcome on mosaicscience.com

Acknowledgements

The stories in this book were originally published by Wellcome on mosaicscience.com. Canbury Press acknowledges the people who worked on these stories for Mosaic, including:

Charlotte Huff, Chrissie Giles, Rob Reddick, Francine Almash, Charlie Hall, Rebecca Guenard, Peta Bell, Kirsty Strawbridge, Penny Bailey, Madeleine Penny, Lowri Daniels, Ian Birrell, Louisa Saunders, Tom Freeman, Neil Steinberg, Holly Cave, Michael Regnier, Jo Marchant, Mun-Keat Looi, Jim Giles, Rhodri Marsden, Louisa Saunders, Mary-Rose Abraham, Emma Young, Christie Wilcox, Katherine Mast, Cameron Bird, Rose George, Cheryl Grant, Liana Aghajanian, Henry Nicholls, Bryn Nelson, Lena Corner, Carl Zimmer, and Andy Extance.